Current Trends in Veterinary Vaccines and Diagnostics

THE EDITORS

Dr. (Mrs.) Anju Nayak is presently working as Professor and Head, Veterinary Microbiology, College of Veterinary Science and Animal Husbandry, Jabalpur (M.P.). She has a vast experience of veterinary profession. She is the recipient of University Gold Medal for undergraduate programme from I.G.K.V.V., Raipur (C.G.). She did her post graduation with specialization in Veterinary Immunology from Indian Veterinary Research Institute. She has pursued her Ph.D. from College of Veterinary Science and A.H., JNKVV, Jabalpur. She had guided 06 PhD, 03 MVSc, one MSc student and had a number of research articles in reputed Journals.

Dr. Vandana Gupta is working in the capacity of Associate Professor in the Department of Veterinary Microbiology, College of Veterinary Science and A.H., JNKVV, Jabalpur. She has done her Masters in Veterinary Microbiology from Anand Agriculture University, Anand. She has pursued her Ph. D from College of Veterinary Science and A.H., NDVSU, Jabalpur. She holds 16 years experience in teaching and Research. She has bagged National awards, travel grants, fellowships and has a number of research and popular articles to her credit.

Dr. Poonam Shakya, Associate Professor, Veterinary Microbiology, NDVSU, Jabalpur, M. P., completed her Bachelor's degree in 2006 and Master's degree in 2008 from Veterinary College, Mhow and grabbed the University Gold Medal. Recently, she has completed her PhD in 2022 and holds an experience in teaching and research of about 10 years. She has a number of scientific publications of national and international repute. She is the recipient of "Fellowship for Training of Young Scientist" by MPCST, Bhopal and Inspiring Lady Veterinarian Award-2021 (Academia Category) by Pashudhan Praharee. Her research interest includes antimicrobial resistance and alternatives to antimicrobials.

Dr. Ajay Rai has been working as Assistant professor in the Department of Veterinary Microbiology, College of Veterinary Science and Animal Husbandry, Jabalpur (M.P) since 2015. He has graduated from College of Veterinary Science and Animal Husbandry, Mhow (M.P.) in 2010. He has completed his Masters in Veterinary Bacteriology from Indian Veterinary Research Institute, Izatnagar, Bareilly (U.P.) in 2012 and PhD from College of Veterinary Science and A.H., JNKVV, Jabalpur in 2022. He holds 10 years' experience in Research and Teaching. He has fellowships and number of research, review and popular articles to his credit.

Current Trends in Veterinary Vaccines and Diagnostics

– *Editors* –

Dr. Anju Nayak

PhD, Professor and Head

Dr. Vandana Gupta

PhD, Associate Professor

Dr. Poonam Shakya

PhD, Associate Professor

Dr. Ajay Rai

PhD, Assistant Professor

College of Veterinary Science and Animal Husbandry,
Jabalpur, Madhya Pradesh

2024

Daya Publishing House®
A Division of
Astral International Pvt. Ltd.
New Delhi – 110002

ISBN 978-93-5461-715-7 (HB)

Published by : **Daya Publishing House®**
A Division of
Astral International Pvt. Ltd.
– ISO 9001:2015 Certified Company –
4736/23, Ground Floor, Ansari Road,
Darya Ganj, New Delhi-110 002
Ph. 011-43549197, 23278134
E-mail: info@astralint.com
Website: www.astralint.com

Printed at : **Replika Press Pvt Ltd.**

Preface

During the COVID-19 pandemic, activities and mobilities among scientific community were limited. To share the knowledge on recent microbiological updates on disease diagnosis and vaccine an online training program was organized. The researchers were very enthusiastic and shared their experience of their laboratory. But due to online limitations the knowledge was shared between limited participants. So, to extend the knowledge among researchers in the form of the present edited book, we are excited to share latest knowledge in area of disease diagnosis, vaccines and its quality control. The present book will be boon for the teachers, scientist as well as students. Every curious researcher will try to keep this book in their shelf for ready reference.

Dr. Anju Nayak

Dr. Vandana Gupta

Dr. Poonam Shakya

Dr. Ajay Rai

Contents

Preface *v*

1. Advances in Vaccine Technologies and their Impact on Managing Emerging Diseases 1
Jagadish Hiremath and H. Rahman

2. Development of Monoclonal Antibody Based Diagnostics 13
S. Bhatia

3. Quality Control and Testing of Veterinary Vaccines 21
M. Arthanari Eswaran

4. Haemorrhagic Septicaemia Vaccines: A Paradigm Shift from Conventional to New Generation Vaccines 31
Bincy Joseph

5. Diagnosis of Aquatic Animal Diseases: Challenges and Future Dimensions 39
K.J. Reshma and T.G. Sumithra

6. Application of Immunohistochemistry in Disease Diagnosis 47
Deepika Lather, Babu Lal Jangir, Vikas Nehra, Gauri Chandratre A. and Gulshan Narang

7. Current Trends in Vaccines and Vaccination of Poultry 57

R. Durairajan

8. Laboratory Biosafety 67

H.V. Murugkar

9. An Overview of Recent Advances in Animal Disease Diagnosis 73

Chintu Ravishankar

10. Nobel Approaches to Detect Biofilm Forming and ESBL Producing Bacteria of Enterobacteriaceae Family from Livestock and Poultry 83

Indranil Samanta

11. Vaccine against Poultry Coccidiosis: A Cup Half Full? 91

Vivek Agrawal, Amit Kumar Jaiswal, Anantrao Kashiram Jayraw, Mukesh Shakya and Nirmal Jamra

12. African Swine Fever in India: Concept and Update 99

T.K. Dutta

13. An Overview of Diagnosis of Mycoplasma Infections 107

Amita Tiwari

14. One Health: A Holistic Approach to Restrict Outbreaks 111

Namita Joshi and R..K. Joshi

15. Current Trends in Diagnosis and Prophylaxis of Canine Morbillivirus (Canine Distemper Virus) 121

K.K. Rajak, M. Bhatt, Kiran, Monu Karki, Vishal Rai, A. Fayaz, A. Kumar, A.K. Yadav, V. Upamanyu and R.P. Singh

16. Recent Trends in Diagnosis of Parvo Virus Infection 133

Shubhangi Warke

17. Recent Update of Footrot in India 139

S. Vijayalakshmi

18. Approaches in Diagnosis of Diseases of Big Cats 145

Nidhi Rajput

19. **Current Status of Bovine Herpesvirus-1 Infections in India** 153

Vipan Kumar, Rinmuanpuii Ralte, Naripjeet Kaur and P.N. Dwivedi

20. **Technology Advances in Diagnostic Microbiology** 157

T.G. Sumithra, K.J. Reshma and V.N. Anusree

21. **Heavy Metal Toxicity in Animals** 165

Amita Dubey, Arya Anil and Yamini Verma

22. **Antibiotic Resistance Patterns of Methicillin-Resistant *Staphylococcus aureus*** 175

B.B. Javia and V.R. Nimavat

23. **Peste-des-Petits-Ruminants: Current Status and Future Perspectives** 183

D. Muthuchelvan, Manisha Medhi, R. Mageswary, R. Ashwini Chaple, S. ChandraSekar, and M.A. Ramakrishnan

24. **Applications of Nanobodies in Diagnosis and Control of Infectious Diseases** 197

Ajit Singh

25. **Control and Elimination of Rabies** 205

Shrikrishna Isloor

Index 209

Chapter 1

Advances in Vaccine Technologies and their Impact on Managing Emerging Diseases

Jagadish Hiremath[1] and H. Rahman[2]

[1]*Senior Scientist, ICAR-National Institute of Veterinary Epidemiology and Disease Informatics (NIVEDI), Bengaluru*
[2]*Regional Representative for South Asia, International Livestock Research Institute, New Delhi*

"Prevention is better than cure"

Desiderius Erasmus, The Dutch philosopher

Introduction

Healthy livestock is the wealth of the nation as it contributes significantly to human health and the country's economy. There are a number of diseases, especially infectious diseases that cause large scale economic loss due to high morbidity and mortality. In India, the combined annual loss due to FMD (moderate outbreaks), HS and Brucella in bovines is estimated to be INR 27545 Crores. Effective disease control and prevention strategies are must to overcome such losses. If understanding disease epidemiology can provide means for disease control and prevention, understanding the nature of pathogens leads to vaccine development. Vaccine is the most effective means of disease control and prevention in human and animals. Vaccines have also resulted in successful eradication of diseases like Rinderpest in bovines and Smallpox in humans. Currently, in India national level vaccination campaigns are carried out to prevent FMD and Brucellosis in cattle, buffalo, sheep, goat and pigs under National Animal Disease Control programme.

Vaccine is an agent specific antigen formulation with or without adjuvant that protects hosts against the disease by inducing specific immune response which lasts for a varied time depending upon nature of the antigen. Body's mechanism to protect against infection involves complex interaction between host and microbe. Multiple organs and cell types communicate between them at different levels to activate the functions that finally safeguard the host against the impact of infection. Traditionally, vaccines were either live attenuated or inactivated which on inoculation produced broad spectrum immune response. Unlike inactivated vaccines, in live vaccines the agent is capable of replicating within the host resulting in strong immune response thereby attaining the desired vaccine effect. However, the live vaccines in immunocompromised hosts can sometimes cause the disease. Hence, the alternate vaccine technologies are emerging that are safe and efficacious. The advancement in molecular biology and biotechnology has led to modern vaccines including viral vectors, nucleic acid-based RNA and DNA vaccines, and virus-like particles.

The antigen component of a vaccine can be live organism (for example, Brucellosis, PPR), killed whole organisms (for example, Hemorrhagic septicemia, FMD), purified proteins from the organism, recombinant proteins, or polysaccharide. The immunogenicity of inactivated vaccines is poor compared to live vaccines and adjuvants are combined to improve the immunogenicity. There are very few adjuvants that are permitted to use with vaccines and research to identify or to develop a potent and safe adjuvant for vaccine formulation is a priority area of research.

The host immune response to the vaccine antigen is analyzed by evaluating the humoral and cell mediated immune response. The innate immune response mediated mainly by antigen presenting cells (Macrophages, dendritic cells), is the first line of the immune system. It presents the vaccine antigen to cells of the adaptive immune system to induce specific immune response with immunological memory. The adaptive immune response has two wings, one mediated by B cells and other by T cells. The antibodies are immunoglobulin proteins secreted by activated B cells that neutralize the pathogens, whereas an activated T cell in addition to its regulatory role, the cytolytic activity clears the infected cells.

Principle of Vaccine Mediated Disease Prevention

A vaccine formulation once inoculated into the host, its ability to induce the immune response and protect against subsequent infection depends on many factors like vaccine antigen, adjuvant, antigen delivery system, route of vaccination, host immunecompetence, number of doses, *etc.* The immune response to vaccine is triggered by antigen that are recognized by pattern recognition receptors of antigen presenting cells (APC) resulting in receptor mediated uptake of protein antigen. The activated APC secrete interleukins and migrate to draining lymph nodes where they present the processed antigen to T lymphocytes. The activated T lymphocytes drive the B cell development in lymph nodes that results in maturation of antibody response to increased antibody affinity and induce secretion of different antibody isotypes. The plasma cells are mature B cells that are short lived but known to produce antibodies in large quantities which increases concentration in serum over a period of up to two weeks. Additionally, the memory cells formed in the process remain in draining lymph nodes of effector sites. Similarly, the CD8+ T cells multiply in large numbers and cause the lysis of infected cells. The body's immune response to first exposure to pathogens is primary response whereas, the second and subsequent response is secondary response.

The immune memory is one of the important features that distinguishes adaptive immunity with innate immunity. The antibody levels and effector CD8+ T cells wane after primary vaccination to a very low level many times even beyond the protection levels. The pool of memory cells formed during the process are critical which decides the strength of immune response in subsequent exposure of pathogens.

Components of Vaccine

Vaccine has different components like antigen, stabilizer, adjuvant, preservatives, and antibiotics (Figure 1.1). Not all the vaccines will have all the listed components, but it depends on the nature of antigen, process of manufacturing, safety norms and source of antigen. Antigen is the primary component that decides the specific immune response for a given pathogen. The other components will assist in either enhancing the vaccine efficacy or will make the vaccine safer. The stabilizer used in the vaccine maintains the vaccine antigen in stable form during storage and most commonly used stabilizers include $MgCl_2$, $MgSO_4$, lactose-sorbitol and sorbitol-gelatine. Adjuvants are added to improve the immune response to vaccine antigen. The Aluminium salts are among the oldest adjuvants that are commonly used both in human and animal vaccines. Preservatives are added to multidose vaccines to prevent bacterial and fungal growth. Thiomersal, Formaldehyde, or Phenol derivatives are some of the preservatives in use. Of the listed, thiomersal is the most commonly used preservative and it has been in use since 1930.

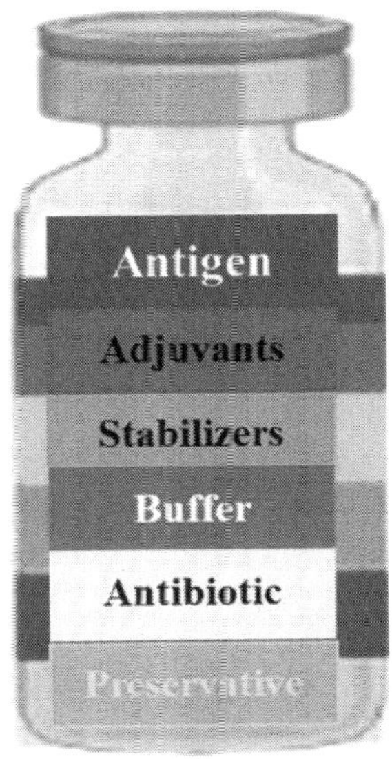

Figure 1.1. Components of Vaccine.

Vaccine Technologies

Technology of making vaccines began way back in the 17th century. Edward Jenner, the founder of vaccinology who inoculated a 13year-old-boy with vaccinia virus (cowpox) and demonstrated immunity to smallpox in 1796. Since then the field of vaccinology is advancing with the goal to produce potent vaccines for all vaccine preventable diseases. Traditionally, live attenuated vaccines and killed vaccines were more preferred vaccine technologies. With recent advancements in molecular biology and biotechnology, several new technologies are developed and evaluated. The following section gives a brief account on various vaccine technologies with their strengths and weaknesses.

Live Attenuated Vaccines

One of the oldest vaccine technologies, available since 1950s, live attenuated vaccines (LAV) are derived from disease-causing pathogens (virus or bacteria) that have been weakened under laboratory conditions by multiple passaging using cell culture or passaging in unnatural hosts (Figure 1.2). They will multiply in a vaccinated individual, but because they are weakened and lost their virulence, they will cause no or very mild disease. There are few live attenuated vaccines that are in use in veterinary practice to name a few are *Brucella abortus* strain S19 and PPR vaccine strain. Live attenuated vaccines stimulate host immune response which is at par with active infection and since the antigen is made available continuously the memory cell produced in good numbers. Though the live attenuated vaccines are potent and produce long lasting immunity they pose safety concerns. The attenuated agent used as a vaccine strain in the host may revert back to virulent form and can show disease symptoms.

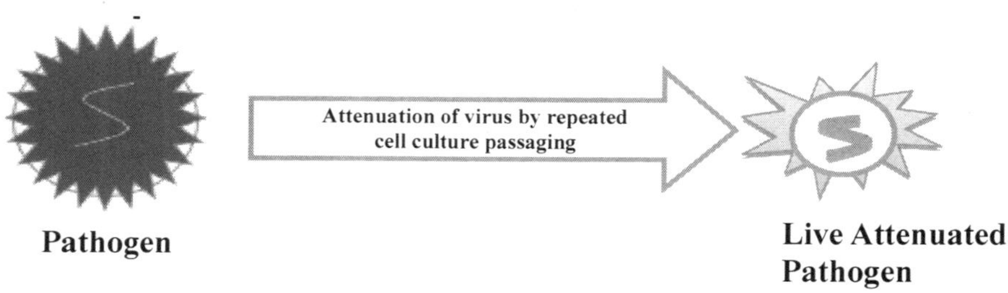

Figure 1.2. Live Attenuated Vaccines.

Killed/Inactivated Vaccines

The killed or inactivated vaccines are those in which the micro-organisms are made completely inactive by chemical or physical process (Figure 1.3). The killed organisms cannot cause disease and hence these vaccines are safe. Immune response by the inactivated whole-cell vaccines is not long lasting and multiple doses of vaccine are required to maintain the protective levels of immunity. The greatest advantage with killed vaccines is that they pose no risk of inducing the disease as they do not contain live components of the pathogen. The killed vaccines used in veterinary practise are against Anthrax, FMD, Haemorrhagic septicaemia, *etc.*

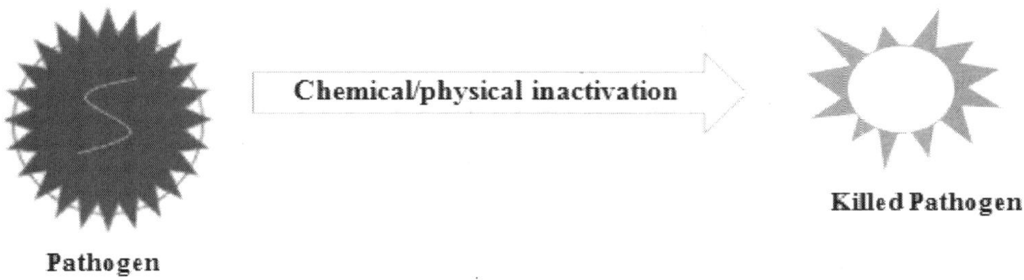

Figure 1.3. Killed/Inactivated Vaccines.

Subunit Vaccines

Subunit vaccines are those that contain one or more antigenic components of a pathogen without having whole of pathogen (Figure 1.4). The antigenic component of the pathogen like lipo-polysaccharide, surface proteins, viral like particles (VLP) and chimeric recombinant proteins, *etc.* are examples of subunit vaccines. Broadly, the subunit vaccines are classified as protein based, polysaccharide based and conjugate based subunit vaccines. The subunit vaccines are safer compared with killed vaccines that have the entire pathogen, but they need adjuvant for inducing good immune response. The spectrum of immune response induced by sub-unit vaccines is narrow compared to killed vaccine and this also affects the immunological memory.

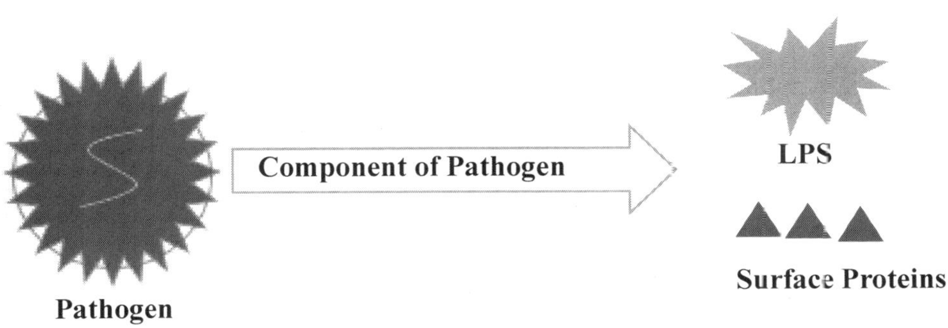

Figure 1.4. Subunit and Conjugate Vaccine

DNA Vaccines

The advancement in nucleic acid vaccines is promising as it has great scalability and has very minimal concerns of safety. DNA vaccines include plasmid DNA with an antigen coding region which upon inoculation is taken up by host cells and inside the cells the antigenic proteins are expressed which further cause the activation of immune response (Figure 1.5). DNA vaccines are capable of inducing humoral and cell mediated immune responses. The antigenic protein expressed inside host cells localizes to the plasma membrane and undergoes post-transcriptional modification resulting in a structure that closely resembles the native structure of an antigen. This is important as it provides the conformational epitope for optimal induction of host immune response. The technology has great potential to scale-up production once the selected plasmid is proven for a particular antigen, it can be easily used for producing desired protein antigen. Hence, under an emergency situation like COVID-19 pandemic, DNA vaccines support rapid vaccine production as they are proven for their safety, production, processing and purification methods for other antigens. Though use of DNA vaccines offers many advantages, one major disadvantage is the long-term persistence of DNA plasmids into the host cells. Studies have shown the presence of plasmid DNA for up to 2 years possesses an additional risk of integration of DNA plasmid with the host genome.

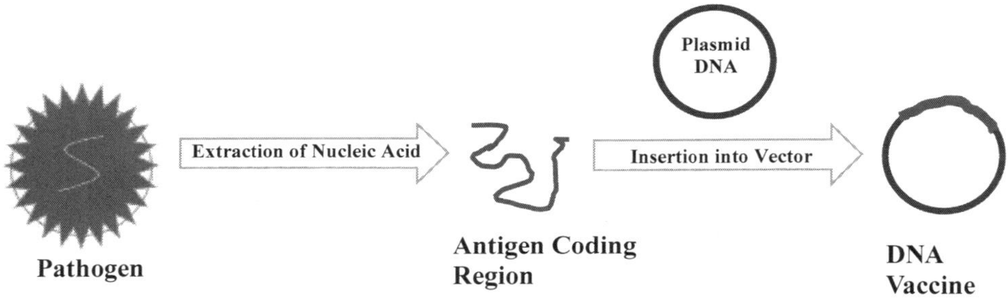

Figure 1.5. Development of DNA Vaccines.

mRNA Vaccines

The mRNA is an intermediated molecule between DNA and protein which carries the coded genetic information necessary for protein synthesis from nucleus to cytoplasm. The process by which mRNA is synthesised is called transcription. The use of mRNA as a vaccine candidate is a recent advancement in the field of vaccinology which is promising technology especially in the era of emerging infectious diseases like COVID-19. Replicating and non-replicating mRNA are two major RNA types that are used as vaccine candidates. Replicating mRNA has viral replication machinery along with an antigen coding region in the mRNA (Figure 1.6). This arrangement facilitates translation of mRNA to produce viral replicating proteins which further multiply the mRNA by the process of replication. The large quantities of mRNA produced is subsequently translated to antigenic protein. Hence, the replicating mRNA vaccines are efficient in inducing host immune response as they produce large quantities of antigenic protein.

The non-replicating mRNA are designed with a sequence of antigen flanked on either side (5′ and 3′) by untranslated regions (Figure 6). The non-replicating RNA are easy to design and small in size resulting in easy synthesis. The immune response induced by these mRNA molecules is specific to antigen protein coded in mRNA. The *in vitro* transcription of plasmid DNA in expression hosts like *E coli* produces non-replicating RNA. Upon purification using HPLC, the impurities are removed and purified mRNA is used for prophylactic or preventive vaccination. Higher the purity of mRNA, better is the yield of antigenic protein, which further influences the strength of host immune response. The critical step in the success of mRNA vaccine is its delivery. The exogenous mRNA has to enter directly into the cell as the extracellular environment may degrade the mRNA. The methods used to deliver mRNA direct injection of naked mRNA, gene gun, *in vivo* electroporation, mRNA complexed with other components and polymer based nanoparticles.

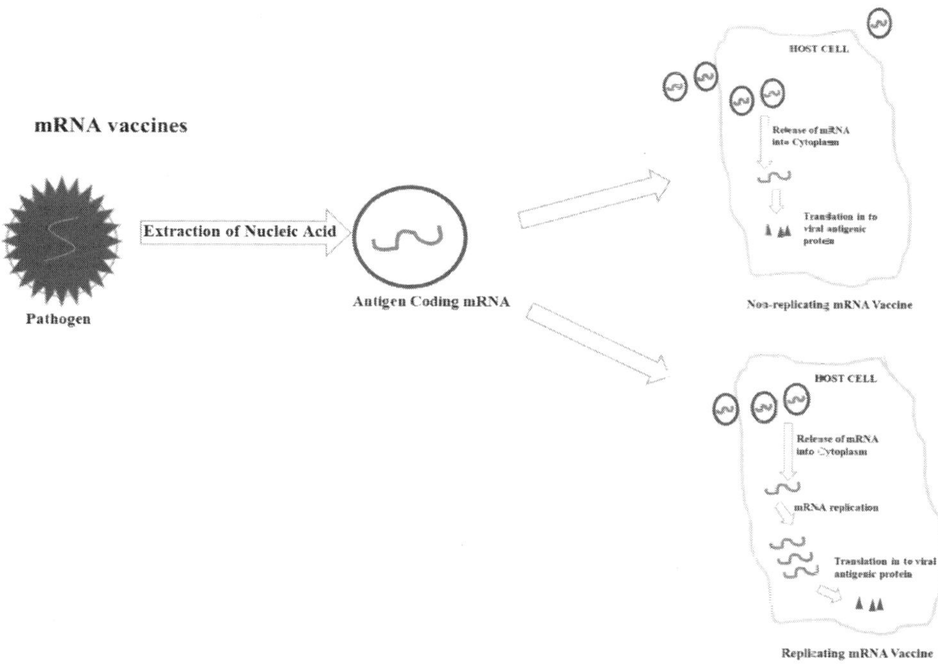

Figure 1.6. mRNA Vaccines.

Adjuvants and their Role in Improving Vaccine Efficacy

Not all types of vaccine formulation are inherently capable to stimulate host immune response optimally. In order to enhance the vaccine mediated host immune response some agents called adjuvants are added to vaccine formulation. Adjuvants are components of vaccine formulation that are added to enhance the immunogenicity of antigens and induce protection against infection. Vaccines made from live-attenuated or inactivated pathogens can elicit robust protective immune responses whereas protein-based vaccines in most cases have limited immunogenicity. Thus, adjuvants are necessary to assist such immunogenic proteins to become effective vaccines by inducing strong and long-lasting protective immune responses. Adjuvant can selectively stimulate either humoral or cell mediated immune response. For example, the aluminium salts which are used for human vaccines since ages are known to induce a strong antibody response but their ability to induce CMI response is poor. Hence, selection of appropriate adjuvant will help to overcome the limitations of vaccine antigen in improving the antigen-specific immune response. Adjuvant benefit in many ways to improve the vaccine effect. The dose of vaccine and number of doses of vaccine are significantly reduced in the presence of appropriate adjuvant. The adjuvants can improve the vaccine efficacy overcoming the host factors like age (infant and elderly) and immunocompromised states. The host immune response is significantly altered by inducing more rapid and long-lasting cell mediated and humoral response.

Advancement in Vaccine Delivery Methods

The effective delivery of vaccine into the host is equally important to attain the desired concentration of antigen in the host system to achieve expected vaccine effect. The emulsions, microparticles, liposomes and immune stimulating complexes are few of the commonly known vaccine delivery systems. The vaccine delivery system using nanoparticles is a recent development which is the most efficient delivery system having an additional adjuvant effect. The advancement in the field of nanotechnology has opened up greater possibility of regulating the size of the antigen with the help of different polymers and formulation strategies, which improved vaccine delivery to the site of effect. In addition to size, nanotechnology offers an opportunity to regulate the shape and surface properties of the entrapped vaccine particles, which significantly contributes to the efficacy of vaccines. In general nanoparticles are used as a delivery system to enhance antigen processing. There are different types of nanoparticles *viz.*, polymeric nanoparticles, inorganic nanoparticles, liposomes, immunostimulating complex, virus like particles, self-assembly proteins, and emulsions. Of all nanoparticles, polymeric nanoparticles being biocompatible and some of them even being biodegradable are studied well. A wide variety of polymers that are either natural (*e.g.* chitosan, collagen, cyclodextrin, pullulan, inulin, alginate and gelatin) or synthetic [*e.g.* polyesters, poly lactic-co-glycolic acid (PLGA), polyacrylates, polylactones and polyanhydrides] were used to prepare nanoparticles with entrapped antigen. Among them, PLGA is widely used in drug and vaccine delivery systems since it is both biocompatible and biodegradable. Importantly, PLGA has been approved by FDA and EMA

(European Medical Agency) for drug and vaccine delivery in humans. There are number of studies where PLGA has been used to carry antigen of different bacterial, viral and protozoal antigens showing improved immune response in each case.

Use of Advanced Vaccine Technologies in Management of Emerging Disease

Currently available vaccine technologies have played a significant role in control and prevention of human and animal diseases. The traditional vaccine technologies like live attenuated vaccines and killed vaccines have led to eradication of some of the animal and human infectious diseases regionally or even globally. The fact that globalization has contributed to significant ecological changes is well known and consequences are observed in the form of emergence of new infectious disease more frequently. With increased frequency of emerging infectious diseases, it has become more important to look for the vaccine technologies that can be employed to produce vaccines at faster pace fulfilling the safety, efficacy, and effectiveness standards. DNA vaccines and mRNA vaccines have many features that make these technologies more suitable during emergency vaccination against emerging diseases like COVID-19. These technologies have been evaluated for their design safety and they have advantage of scalability within stipulated time. It is the antigenic region inserted into plasmid vector or coded in mRNA needs to be changed with changing pathogen, but basic framework remains same. Hence, DNA vaccine and mRNA are promising technologies that have been used to produce COVID-19 vaccines and have proved their ability in emergency vaccine production.

Challenges in Vaccine Development

Vaccine development is long process starting from design of vaccine, validation, trial, approval to manufacture, supply and finally vaccination of target population. At each step there are number of challenges of which safety is a major concern. With increasing number of emerging infectious diseases and their pandemic potential it has become much more challenging to develop a vaccine within stipulated timeline. An average time period required by a vaccine to reach market from design to approval for manufacturing is around ten years for human vaccines with approximate cost of 500 million USD. Technically, the RNA viruses pose greater challenge due to their high rate of mutation leading to antigenic changes in virus making the current vaccine ineffective. Hence, a strong national level surveillance mechanism is required to captures the changes in circulating RNA viruses and use such information to modify the vaccine strains. The challenges are much more with resource limited countries. Apart from technical challenges the other challenges include infrastructure, trained manpower, delay in regulatory approvals, lack of foolproof guidelines makes vaccine development a daunting process in such countries.

Conclusion

Vaccine is one single human invention that has brought great health to masses. In veterinary field, the economic loss due to infectious disease was significantly reduced by vaccination programs. Number of vaccine technologies evolved over the period of time matching the advances in allied disciplines. Every technology has its own advantages and disadvantages. Choosing the right technology for given agent is necessary. Globally, with changing ecology the human and animal population are witnessing new and emerging infectious diseases which challenge the existing vaccine technologies. The advancement in vaccine development is regular process with aim to produce efficacious and effective vaccine.

References

1. Lee S, Nguyen MT. (2015) Recent advances of vaccine adjuvants for infectious diseases. *Immune Netw.*; 15(2): 51-57.
2. Pardi, N., Hogan, M., Porter, F. *et al.* (2018) mRNA vaccines — a new era in vaccinology. *Nat Rev Drug Discov* 17, 261–279.https: //doi.org/10.1038/nrd.2017.243
3. Rauch S, Jasny E, Schmidt KE, Petsch B. (2018). New Vaccine Technologies to Combat Outbreak Situations. *Front Immunol.* 19; 9: 1963.
4. Saroja Ch, Lakshmi P, Bhaskaran S. (2011) Recent trends in vaccine delivery systems: A review. *Int J Pharm Investig.* 1(2): 64-74.
5. Heaton PM. (2020) Challenges of Developing Novel Vaccines with Particular Global Health Importance. *Front Immunol.* Oct 14; 11: 517290.
6. World economic forum website (https: //www.weforum.org/agenda/2020/06/vaccine-development-barriers-coronavirus/)
7. World Health Organization. *Vaccine Technologies.* Available online (https: //www.who.int)
8. Brisse M, Vrba SM, Kirk N, Liang Y, Ly H. (2020) Emerging Concepts and Technologies in Vaccine Development. *Front Immunol.* Sep 30; 11: 583077
9. Trovato M, Sartorius R, D'Apice L, Manco R, De Berardinis P. (2020) Viral Emerging Diseases: Challenges in Developing Vaccination Strategies. *Front Immunol.* Sep 3; 11: 2130.
10. Pollard AJ, Bijker EM. (2021) A guide to vaccinology: from basic principles to new developments. *Nat Rev Immunol.* Feb; 21(2): 83-100.
11. G G, B GK, A K, Hegde R, Kumar N, Prabhakaran K, Wadhwan VM, Kakker N, Lokhande T, Sharma K, Kanani A, Limaye, K N, Pn A, De AK, Khan TA, Misri J, Dash BB, Pattnaik B, Rahman, H. (2021). Foot and Mouth Disease (FMD) incidence in cattle and buffaloes and its associated farm-level economic costs in endemic India. Prev Vet Med. 2021 May; 190: 105318.

12. Singh B, ShivPrasad, Med Ram Verma and D.K.Sinha (2014). Estimation of economic losses due to Haemorrhagic Septicaemia in cattle and buffaloes in India. Agricultural *Economics Research Review* 27(2): 271-279

13. Singh BB, NK Dhand and JPS Gill (2015). Economic losses occurring due to brucellosis in indian livestock populations. *Prev vet Med* 119(3-4): 211-5.

Chapter 2

Development of Monoclonal Antibody Based Diagnostics

S. Bhatia

Principal Scientist, ICAR-National Institute of High Security Animal Diseases, Bhopal – 462 022, Madhya Pradesh

Introduction

In 1975 George Kohler, a post doctoral student at Cambridge University in Dr. Cesar Milstein's laboratory first produced a hybridoma by fusing normal lymphocyte and myeloma cells using sendai virus. The hybridoma cells could be cloned and single cells could be selected and grown. As a result, the products derived from a single clone of cells were called as **monoclonal antibodies**. This method of producing monoclonal antibodies developed by Kohler and Milstein has since been refined with many variations.

Over the past few decades, monoclonal antibodies (Mabs) have become the preferred source of antibodies for diagnosis of infectious diseases. They are pure and absolutely specific for individual antigenic determinants and can be produced in unlimited amounts. As a result, they are rapidly being incorporated in clinical diagnostic techniques where large amount of specific antibody of consistent quality are required. Though Mabs are useful in many areas of biological research, they have completely revolutionized the area of clinical diagnosis.

At present, there are many diagnostic kits available commercially which make use of monoclonal antibodies. Some of the important animal diseases for which Mab based kits are available are Foot and Mouth Disease (FMD), Rinderpest (RP), Bovine Viral Diarrhea (BVD), Malignant Catarrhal Fever (MCF), Salmonellosis, Brucellosis *etc.* Such diagnostic kits based on monoclonal antibodies provide specific and sensitive diagnosis for many other infectious diseases. Keeping in view of the usefulness of monoclonal antibodies in diagnosis, the production and characterization of Mabs to bovine immunodeficiency virus (BIV) gag protein has been aimed in this study.

Production of Monoclonal Antibodies against Specific Antigen

Production of murine monoclonal antibodies against the desired specific antigen of the pathogen is done by carrying out a series of procedures described below with minor modifications of protocol suggested in "Monoclonal Antibodies" (ed.1992) by J. H. Peters and Baumgarten.

Immunization of Balb/c Mice

Female Balb/c mice of 6-8 weeks of age are immunized with purified antigenic (50 µg) along with the adjuvant (FCA) in 1:1 ratio on day 1 and with FIA on day 14, 21, 28 via intra-peritoneal route and on day 35 via intravenous route.

FCA = Freund's Complete Adjuvant

FIA = Freund's Incomplete Adjuvant

The serum samples are collected from each mouse before the first injection and after the last intraperitoneal injection. The test bleed sera may be titrated by indirect ELISA for presence of antibodies specific to the desired antigen.

Collection of Peritoneal Macrophages (Feeder cells)

The peritoneal macrophages from un-immunized mouse are used as feeder cells for growth of hybridoma cells. The feeder cells phagocytise dead cells and release various interleukins, which help in growth of hybridoma cells. The feeder cells are seeded in 96 well plates 4-5 days prior to seeding of the hybridoma cells to support the growth of hybridoma.

1. Five Balb/c mice (non-immunized) are killed by cervical dislocation and fixed on a sterilized cushion. Abdominal region is sterilized with 70 per cent ethanol and dissected open without puncturing the abdominal muscles.
2. Ten ml of medium (RPMI-1640) is inoculated in the abdominal cavity. Abdomen is then massaged gently and the medium was sucked out slowly.
3. The cells collected from all the five mice are pooled and centrifuged at 1200 rpm for 10 min.

4. The supernatant is discarded and the pellet is re-suspended in 2 ml of growth medium (RPMI-1640 with 20 per cent FBS and 1x HAT).
5. The cells are counted and the concentration is adjusted to 10,000 cells/ml by adding more growth medium.
6. With a multi-channel pipette, 50 µl of cell suspension is dispensed in each well of 96-well tissue culture plates and incubated at 37°C for 4 days or till the fusion is done.

Collection of Spleen and Preparation of Spleen Cells for Fusion

1. One of the mice immunized with the antigen is sacrificed by cervical dislocation and its abdomen is cut open with sterile scissors and forceps. The spleen is collected in sterile RPMI-1640 medium in a tube.
2. The collected spleen is washed 3 times with medium and 5 ml of RPMI-1640 medium is injected into spleen to disrupt the stroma. The capsule of spleen is removed and it is pressed with syringe plunger to collect cells.
3. The collected cell suspension is centrifuged at 1200 rpm for 10 min at 25°C. The supernatant is discarded and 1 ml of RBC lysis buffer is added to the cell pellet.
4. 15 ml of RPMI-1640 medium is added and centrifuged at 1200 rpm for 7 min.
5. Steps 3 to 4 are repeated to achieve complete lysis of RBCs. The cells are re-suspended in 10 ml of fusion medium. Counting of cells is done using Neubauer's chamber (Haemocytometer) and the cells are kept at room temperature till myeloma cells are ready.

Fusion of Myeloma and Spleen Cells

1. The spleen cells and myeloma cells are mixed at 2.5: 1 ratio in a 15 ml tube. The volume is made up to 10 ml with fusion medium.
2. The mix of cells is washed twice (1200 rpm for 10 min) in fusion medium.
3. One ml of buffered PEG 1400 (ready-to-use) at 37°C is added gradually over 1 min with slow rotation of tube.
4. The tube is kept in 37°C water bath for 90 secs.
5. One ml of fusion medium is added gradually over 1 min.
6. After pause of 1 min, 2 ml of fusion medium is added over 1 min with gentle shaking and 15 ml of fusion medium is added gradually.
7. The cells are centrifuged at 1200 rpm for 10 min and the pellet is re-suspended in 120 ml of growth medium with 1x HAT.

8. With multi-channel pipette hybridoma cells are dispensed in twelve 96-well flat bottom tissue culture plates (100 µl/well).
9. The plates are then incubated at 37°C in humidified incubator with 5 per cent CO_2. Few wells are randomly checked under inverted microscope for presence of hybrid cells.

Cultivation of Hybridomas

1. The hybridoma plates are left undisturbed for 48 hours and then checked for presence of bacterial contamination if any prior to further work.
2. On 4th day of incubation, 50 µl of medium from each well is removed and 50 µl of fresh medium is added in each well.
3. On 7th day 50 µl of medium (culture supernatant) is taken from each well and tested in indirect ELISA for presence of antibodies.
4. The cultures (wells) found positive in ELISA were transferred to 24-well plate.
5. The second screening by ELISA is done on 13-15th day and the positive cultures are transferred to 24-well plate.
6. Sub-cloning is done to isolate cells specific for antigen from a mixture of population of hybridoma cells in the primary clone since primary hybridoma clones might contain other hybridoma cells not producing specific antibodies.

Diagnostic Application of Monoclonal Antibody

The application of monoclonal antibodies in diagnosis is by far the most advanced, especially for tests that are performed on body fluids such as blood and urine. Monoclonal antibodies can be used to detect the presence of antigens in different techniques which include ELISA, Western blot, Lateral flow cytometry, immunohistochemistry, radioimmunology assay, electron microscopy, fluorescence microscopy and confocal microscopy. Additionally, monoclonal antibodies are used for the detection of antigen as well as antibodies against bacterial diseases of livestock; for detection of antigen as well as antibodies against fungal diseases of livestock; for detection of antigen as well as antibodies against viral diseases of livestock and diagnosis of parasitic diseases.

The basic mechanism of a MAb is the same as an antibody produced by the body. However, when MAbs are used in the diagnosis and treatment of diseases, certain substances are often added to them to give them their therapeutic and diagnostic characteristics. They can also be used on their own to block or encourage certain responses from the immune system. In diagnosis, radioactive markers are attached to them to locate a certain kind of cell within the body.

Immunohistochemistry

Immunohistochemistry (IHC) or immunocytochemistry is a method for the localization of specific antigens in tissues or cells using antibody probes. Because it provides detailed and specific information on the expression of cell markers, the IHC technique has wide application in the identification and classification of cell types, with particular value in diagnostic surgical pathology for infectious disease and tumor diagnosis. A milestone in the development of IHC has been reached with its expanding use of routinely processed, paraffin-embedded tissue sections, an achievement made possible by a series of technical accomplishments that include increased sensitivity of detection, monoclonal antibody (MAb) technology, and antigen retrieval methods to recover antigenicity lost during tissue processing. In the past 10 yr, IHC has made a significant impact on diagnostic surgical pathology, becoming one of the most useful auxiliary diagnostic procedures for the visualization of specific antigens in tissue sections and cell smears. In a large proportion of difficult or ambiguous cases of histopathologic diagnosis, IHC can enhance diagnostic accuracy, particularly in the area of neoplastic processes. Despite problems of insufficient quality control and a lack of standardization of many IHC reagents, the IHC technique is now commonly used as a daily adjunct to the traditional staining methods of diagnostic surgical pathology. With the help of monoclonal antibodies, immunohistochemical procedures such as immunoperoxidase have been improved and used as important tool for the detection of viral diseases.

Enzyme-linked Immunosorbent Assay (ELISA)

Several immunological assays have been developed for the diagnosis of infectious diseases which were either through detection of antigen or antibodies in sera of infected animal or individuals. Competitive ELISA assays based on mAbs not only allow specific detection but also aid accurate quantification of antibodies to human and animal viruses. For example, competitive ELISA developed for bovine immunodeficiency virus (BIV) (Bhatia *et al.*, 2008), bovine viral diarrhea (BVD) (Bhatia *et al.*, 2008) and peste des petits ruminants (PPR) (Singh *et al.*, 2004). Similar c-ELISA assay was found useful for the evaluation of H5 type influenza virus in samples that are difficult to be evaluated by haemaglutination inhibition test. Sandwich ELISAs for detection of antigen also use mAbs as detecting or capture antibodies and are widely used in this format of ELISA. A number of MAbs based ELISA kits are available for sero-diagnosis and agent detection of several livestock and poultry diseases.

Western Immunoblotting

Since the inception of the protocol for protein transfer from an electrophoresed gel to a membrane, protein blotting has evolved greatly. Western blotting analysis can detect one protein in a solution that contains any number of proteins and giving the protein information, and it is widely used in protein detection. It is a method in molecular biology/biochemistry/ immunogenetics to detect protein in a given sample of tissue homogenate or extract (Yang and Ma, 2009), which is normally used with high antibody directed against a desired protein (antigen).

Lateral Flow Assays (LFA)

The lateral flow assay (LFA) is a paper-based platform for the detection and quantification of analytes in complex mixtures, where the sample is placed on a test device and the results are displayed within 5–30 min. Low development costs and ease of production of LFAs have resulted in the expansion of its applications to multiple fields in which rapid tests are required. LFA-based tests are widely used in hospitals, physician's offices and clinical laboratories for the qualitative and quantitative detection of specific antigens and antibodies, as well as products of gene amplification. Monoclonal antibodies in unconjugated or conjugated forms are integral part of LFAs and provide specificity to these assays. The unique and remarkable properties of LFAs have contributed to the detection of disease biomarkers and infectious agents in medicine, agriculture, food and environmental safety. Although the principle of the method has remained unchanged for decades, there have been continuous improvements of LFA techniques leading to increased sensitivity and reproducibility, and the simultaneous detection of several analytes. Importantly, these assays can now be effectively performed outside the laboratory, providing great advantages for use in developing countries and at the point-of-care, whether in the field or in more traditional clinical settings.

Immunofluorescent Antibody Test (IFAT)

The immunofluorescence test is described as a useful technique in the immunological diagnosis of strongyloidosis. The method is used to detect antibody present in the serum of patients through binding to surface antigens or within the parasite. The reaction is read in a fluorescence microscope after adding an anti-human Ig antibody conjugated to a fluorochrome. This technique (IFAT) has the advantage of providing a quantitative result by the accurate determination of antibody titer. This factor is particularly useful for therapeutic evaluation. However, it is a more complex technique in relation to other serological methods and requires skilled and trained technical personnel for both the antigen preparation and reading the slides.

Flow Cytometry

Flow cytometry is a method that principally measure optical and fluorescence characteristics of a single cell or particle of a nuclei, microorganism or chromosome preparation. It gives more information about infectious agents than any other available method. While cytometry equipment remains comparatively expensive, there still exist expanding understandings that use of these machines is necessary in endemic populations at or near points of care. The important signals provided about cells by cytometers cannot be neglected even because of the speed at which results are obtained. This has also made cytometry to become of particular significance because it overcomes the shortcomings associated with the non-cytometric methods.

Conclusion

In conclusion, advancements in biotechnology have led to a large scale production of monoclonal antibodies for use in many diagnostic assays. These assays are frequently employed either for the detection of an immunogenic agent or any of its antigenic component. Future modifications of these assays and their application will no doubt help to ease early detection and diagnosis of emerging infections for proper control and preventions.

References

1. Bhatia S., Sood R., Bhatia, A. K., Pattnaik, B. and Pradhan H. K. (2008) Development of a capsid based competitive inhibition enzyme linked immunosorbent assay for detection of bovine immunodeficiency virus antibodies in cattle and buffalo serum. J. Virol. Methods 148: 218-225.
2. Bhatia S., Sood R., Mishra, N., Pattnaik, B. and Pradhan H. K. (2008) Development and evaluation of a MAb based competitive-ELISA using helicase domain of NS3 protein for sero-diagnosis of bovine viral diarrhea in cattle and buffaloes. Res. Vet. Sci. 85 (1): 39-45.
3. Singh RP, Sreenivasa BP, Dhar P, Shah LC, Bandyopadhyay SK. Development of monoclonal antibody based competitive-ELISA for detection and titration of antibodies to *peste des petits ruminants* virus. Vet Microbiol. 2004; 98: 3–15.
4. Yang and Ma (2009). Western Blotting and ELISA Techniques. Researcher, 2009; 1(2): 67-86.

Chapter 3

Quality Control and Testing of Veterinary Vaccines

M. Arthanari Eswaran

Assistant Professor, Department of Veterinary Microbiology,
Veterinary College and Research Institute, Udumalpet - 642126, Tamil Nadu
e-mail: drarthu@gmail.com, arthanarieswaran.m@tanuvas.ac.in

The veterinary biologicals are the backbone of animal industry in India. The development and evaluation of veterinary biological products in India was initiated when the Imperial Bacteriological Laboratory was established at Poona in 1890 under the recommendation of Indian cattle plague commission. In 1932 a committee was constituted by the Government of India to standardize veterinary biological products to have a national quality standard and uniform production procedure. The veterinary biological products came under the purview of the drugs and cosmetics act, 1940 in 1969.

The potent immunobiologicals including diagnostic antigens play a vital role in providing effective animal health coverage. Judicious application of veterinary control measures and use of biological products enable us to reduce the incidence of various infectious diseases such as FMD, black quarter, brucellosis, African horse sickness, fowl pox and eradicated Rinderpest.

At present, there are many state biological production units engaged in the production of several important biologicals and give regular training programmes in the technique of manufacture and standardization of veterinary biologicals to the state veterinarians.

All the vaccines need to be tested strictly as per the pharmaceutical monographs and also to be exported as per the guidelines of international monographs. Following is the list of Indian and international guidelines specially emphasizing the testing of veterinary vaccines.

- Indian Pharmacopoeia, 2014
- British Pharmacopoeia, 2016
- European Pharmacopoeia, 8th edition
- US Pharmacopoeia, 2016
- Japanese Pharmacopoeia, 17th edition
- Title 9 of Code of Federal Regulations
- OIE manual for vaccines, 2015

Quality Controls in Vaccine Production

As per OIE, the Quality control is concerned with sampling, specifications and testing as well as the organization, documentation and release procedures for vaccines to market. Quality control ensures the necessary and relevant tests are carried out, and that materials and equipment are not released for use, nor products released for sale or supply, until their quality has been judged satisfactory. Quality control is not confined to laboratory operations, but must be involved in all decisions that may concern the quality of the product. The independence of quality control from production is considered fundamental to the satisfactory operation of quality control.

A reliable supply of pure, safe, potent, and effective vaccines is essential for maintenance of animal health and the successful operation of animal health programmes. Immunization of animals with high quality vaccines is the primary means of control for many animal diseases. In other cases, vaccines are used in conjunction with national disease control or eradication programmes.

The requirements and procedures described here are intended to be general in nature and to be consistent with published standards that are generally available for guidance in the production of veterinary vaccines (OIE Terrestrial Manual, 2015).

Properties of Vaccine

The consistent production of pure, safe, potent, and efficacious vaccines requires quality assurance procedures to ensure the uniformity and consistency of the production process. As production processes for vaccines provide a great opportunity for variability, care must be taken to control variability to the greatest extent possible, preferably using validated procedures, and to protect the product from contamination through all stages of production. Consistent product quality (batch-to-batch uniformity) must be built in at each stage. Final product testing is used as a check to verify that the controls on the production procedures have remained intact and that the released product meets the specification previously agreed with the licensing authority.

Quality Control of Vaccines

Quality control measures used in manufacturing of Vaccines

1. Testing of raw and packaging materials
2. Testing of in-process materials
3. Testing of final product

1. Raw Materials

SPF eggs should be tested for weight of the eggs, quality of the egg shell *e.g.* porosity and thickness of shell, fertility of the eggs, use egg surface swab test for sterility testing and egg incubator environment monitoring.

Cell line testing for extraneous agents especially BVD, IBRT, Mycoplasma and TSE

Chemicals testing Description as per I.P./B.P./E.P./U.S.P

Testing of water used in manufacturing -Appearance, Colour, Odour, Taste, pH, Conductivity, TOC, Acidity and alkalinity, Ammonium, calcium, magnesium, heavy metal, chloride, nitrate and sulphate test. Microbial testing of water for total bacterial and coliform counting. LAL test for endotoxins produced by Gram negative bacteria

Packing materials-Standard size, shape, materials, toxicity

2. In Process Testing

Sterility testing of media, stabilizer, adjuvant,

Bulk Sterility testing of inoculums

Environment monitoring through air sampler

Finger dab, Contact plate testing, Settle plate exposure during critical activities.

a. Live Antigen Testing

Sterility test

Titration of virus by EID 50 or TCID 50

Titration of bacteria by CFU

Detection of mycoplasma contamination

b. Inactivated Antigen Testing

Inactivation Study Sterility test and test for residual formaldehyde

3. Final Product Testing

Monitoring the quality at every stage is essential in the development and production of any vaccine. Hence, before issuing to the field use, the vaccine must be subjected to the following quality control tests. (A) Sterility test (B) Safety test (C) Efficacy test (D) Potency test. These tests are normally undertaken by the separate independent authority which does not belong to the vaccine production units.

a. Final Product-Live Vaccine

1. Identification test/Purity test- detection of contamination
2. Sterility test
3. Titer estimation by CFU or EID 50 or TCID 50
4. Safety test
5. Potency test
6. Moisture determination
7. Detection of mycoplasma contamination/detection of extraneous pathogens by chicken embryo inoculation

b. Final Product-Inactivated Vaccines

1. Sterility test
2. Monitoring for extraneous antigen in SPF chicks
3. Safety test
4. Potency test
5. Serology test
6. Formaldehyde estimation
7. Emulsion test

A. Sterility Test

The vaccine must be sterile if it is a killed vaccine. This is to assess whether the inactivation is properly done and no viable organisms are present in the vaccine. Live viral vaccines must be bacteriologically and mycologically sterile. This should be carried out by inoculation of the vaccine in the following media. (a) Fluid Thioglycocolate medium (b) Soybean Casein Digest medium/ Saboraud's broth. The sampling must be, not less than 10 per cent of the lot from the bottled or filled up containers. Additional samples may be taken, if repetition of sterility test is warranted.

B. Safety Test

It is essential that a vaccine be safe and not induce an adverse reaction or signs of disease in the vaccinated animal or human being who is to handle the vaccine.

This test is carried out to check:

(i) Whether the inactivation is complete

(ii) The effect of residual virulence in case of live vaccine

(iii) The presence of other pathogenic viruses in case of live viral vaccine.

Safety is determined in the species for which the vaccine is intended. To obtain a reliable conclusion, the test is carried out in the most susceptible species and at the minimum age of recommended vaccination.

The different types of safety tests are given below.

(i) Safety of administration of one dose.

(ii) Safety of one administration of an overdose

(iii) Safety of the repeated administration of one dose

(iv) Examination of reproductive performance

(v) Examination of immunological functions.

In the case of adjuvanted vaccines, it is necessary to study the effect of the vaccine at the injection site, (*e.g.*) residues, tissue lesions *etc.* and their duration.

a. Live Vaccine

Vaccines are inoculated in up to 10 doses in birds/animals of age and route prescribed in pharmacopoeia. Animals are observed for up to 21 days. No clinical signs and adverse effects attributable to vaccine strain should be observed in the period

b. Inactivated Vaccine

Vaccines are inoculated in up to 2 doses in birds/animals of age and route prescribed in pharmacopoeia. Animals are observed for up to 21 days. No clinical signs and adverse effects attributable to vaccine strain should be observed in the period

The safety studies carried out in the laboratory may have to be complemented by field studies. The studies are of importance because the laboratory studies are always carried out in a limited number of animals. The safety of the product will, in any case, be monitored in the post-marketing surveillance system, also called pharmacovigilance.

C. Efficacy Test

Efficacy can be defined as the ability of the vaccine to give protection against the adverse effects of the infection to the vaccinated animal itself, or the ability to induce antibodies in the vaccinated animal that are transferred to the offspring, and so provide protection to the newborn animals. Protection is not necessarily defined as protection against infection but also as protection against clinical disease, loss of production *etc.*

The efficacy of the vaccine is demonstrated for each species for which the vaccine is intended and for each recommended route of administration. The dose of vaccine used in these trials is the normal recommended dose.

Since the efficacy of the vaccine depends on its potency, it is necessary to use a vaccine with a potency not exceeding the potency guaranteed at the end of shelf-life. The efficacy of the vaccine is also examined in accordance with the recommended schedule of vaccination, (*e.g.*) use of priming with live vaccine and booster with inactivated vaccine.

Duration of Efficacy

Vaccination will not provide a life-long immunity and, consequently, the animals have to be revaccinated. The recommended period for revaccination should be mentioned in the directions for use of the product.

Studies to determine the duration of immunity may take up to a period of one year or even upto 2-3 year studies in the case of some rabies vaccines. At the end of the period, the immunity of the animals must be determined. This can be done by challenge or by serological tests, provided there is a correlation between antibody titre and protection.

Vaccine Assessment

To assess the efficacy of a vaccine, animals must first be vaccinated and then challenged with the same virulent organism. The percentage of vaccinated animals and non vaccinated animals (control), that survive this challenge are measured. The true efficacy of a vaccine, called the **preventable fraction (PF)**, is calculated thus:

$$PF = \frac{(\text{Per cent of controls dying} - \text{Per cent of vaccinates dying})}{\text{Per cent of controls dying}}$$

Good, effective vaccines should have a PF of at least 80 per cent. Obviously, less effective vaccines are acceptable if no better is available.

D. Potency Testing

It is essential to guarantee that each and every batch of vaccine will have its intended effect. This is guaranteed by a manufacturing process of consistent quality by the application of Good Manufacturing Practice, and checked by means of the potency test.

Live Vaccine

One dose of vaccine is inoculated in birds/animals of age and route prescribed in pharmacopoeia. Birds/animals are challenged as prescribed in pharmacopoeia and observed for up to 21 days. Vaccinated animals should be protected from the disease and control animals should develop disease during observation period.

Inactivated Vaccine

One dose of vaccine is inoculated in birds/animals of age and route prescribed in pharmacopoeia. Birds/animals are challenged as prescribed in pharmacopoeia and observed for up to 21 days. Vaccinated animals should be protected from the disease and control animals should develop disease during observation period. For many vaccines, serology test is also performed after collecting blood samples from the vaccinated birds. The different types of potency tests are given below.

a. Live vaccines: determination of the number of live organisms: titration; direct count.
b. Inactivated vaccines
   (i) *In vitro* tests : quantitative test of the antigen: determination of number of organism before inactivation; determination of antigen content by ELISA, SRD test, *etc.*
   (ii) *In vivo* tests: challenge test in target species
   challenge test in non-target species
   antibody response in target species
   antibody response in non-target species.
c. Serological Test

Serology test is also performed after collecting blood samples from the vaccinated birds for many of the inactivated vaccines After collecting blood samples post vaccination as per prescribed days in pharmacopoeia, serum is separated Haemagglutination-Inhibition, Agar Gel Precipitation or ELISA tests are performed to estimate specific antibodies produced against the vaccine antigen. Antibody titre is measured for passing limit mentioned in guidelines.

E. Stability Tests

Stability studies (based on an acceptable potency test) are required to establish the validity of the expiry date that appears on the product package. The use of accelerated stability tests to determine a provisional expiry date for products, *e.g.* incubating at 37°C for 1 week for each year of dating. Such estimates must be confirmed by periodic real-time potency tests on at least three different batches/serials through the period of time indicated by the expiry date, and 3–6 months beyond. For products containing viable organisms, testing should be done at release and at the approximate expiry date until a statistically valid record has been established. For non-viable products, each batch/serial presented for licensing is tested at release and at periodic intervals through, or past, the requested expiry date.

Formaldehyde Estimation

Residual/Free formaldehyde of inactivated vaccines should not be more than 0.05 per cent w/v

Qualities of Good Emulsion

A stable emulsion will show homogeneous fine particles and Brownian movement

Moisture Content

Moisture percentage of live vaccines should not be more than 3.0 per cent w/w

Field Trials

All veterinary biological products administered to animals should be tested for safety and, if possible, for efficacy in the field, using good clinical practice, before being authorised for general use. Field studies are designed to demonstrate efficacy under working conditions and to detect unexpected reactions, including mortality that may not have been observed during the development of the product. In cases where the disease does not induce clear clinical symptoms, other criteria to demonstrate protection have to be used (*e.g.*) growth, post-mortem examinations.

Additional Requirements for Live rDNA Products

The release of live rDNA microorganisms (Categories II and III) for field testing or general distribution as an approved or licensed product may have a significant effect on the quality of the human and animal environment. Before release is authorized, the manufacturers of the vaccine should conduct a risk assessment to evaluate the impact on the human and animal environment.

The topics discussed should include:

i) The characteristics of the vaccine organism,
ii) Human health risks,
iii) Animal health risks for both target and non-target animals,
iv) Persistence in the environment, and increase in virulence.

Documentation of the Manufacturing Process and Record Keeping

A detailed Outline of Production, a series of SOPs, or other documents should also be prepared to describe the protocol for the manufacture and testing of each product produced in an establishment. Criteria and standards for source materials should be clearly and accurately documented. Guidelines for the preparation of such documents for veterinary vaccines are published by competent control authorities. This documentation is intended to define the product and to establish its specifications and standards. It should serve, along with the blueprints and blueprint legends (or production plan and SOPs), as a uniform and consistent method of producing the product that should be followed in the preparation of each batch/serial (one master batch record for each product).

References

1. British Pharmacopoeia, 2016.
2. European Pharmacopoeia, 8 th edition.
3. Glen Gifford, Pawan Agrawal, Donna Hutchings, and Oksana Yarosh, 2011. Overview of current regulatory requirements and accepted alternatives. Procedia in Vaccinology 5: 236 – 247.
4. Indian Pharmacopeia 2010
5. Principles of Veterinary Vaccine Production 2015, OIE Terrestrial Manual, P.No. 90-104.
6. Title 9 of Code of Federal Regulations (CFR), United States Pharmacopeia.
7. http: //www.oie.int/en/international-standard-setting/terrestrial-manual/access-online/Veterinary vaccine post-licensing safety testing.

Chapter 4

Haemorrhagic Septicaemia Vaccines: A Paradigm Shift from Conventional to New Generation Vaccines

Bincy Joseph

Assistant Professor, Department of Microbiology and Biotechnology, Post Graduate Institute of Veterinary Education and Research, Jaipur, Rajasthan

Haemorrhagic Septicaemia (HS) is a major bacterial disease of cattle and buffaloes. It is characterized by an acute and highly fatal septicaemia with high mortality rate (≈ 60 per cent) and is prevalent throughout the country. Genetically, buffaloes are more susceptible to it than cattle and young animals are more prone to this than adults (FAO, 1991). There is a sharp drop in milk yield of the affected animals. It has emerged as a disease of considerable economic importance in India where cattle and buffaloes are large in number and are vital for milk production and draught power. It is the second most reported disease in India during 1991 to 2010 (NADRES- National Animal Diseases Referral Expert System) and is the cause of maximum number of the reported deaths. Around 97 per cent of the HS outbreaks reported by NADRES are in large ruminants (Gajendragad and Uma, 2012).

During 2007 to 2011, the average number of HS infected cases reported by the Department of Animal Husbandry, Dairying and Fisheries, Government of India, were 2416 in cattle and 825 in buffaloes. The highest cases of cattle infection were reported from the eastern region (43.71 per cent), followed by western (27.06 per cent), southern (22.42 per cent) and northern (6.81 per cent) regions. The incidence pattern of HS in buffaloes was different; it was highest in southern region (69.93 per cent), followed by northern (20.97 per cent), western (8.51 per cent) and eastern (0.58 per cent) regions. In terms of the period, the effect of HS was observed throughout the year. Season wise, the HS infection (number of cases) in cattle was more during rainy season (40.85 per cent), followed by summer (30.6 per cent) and winter (28.55 per cent) seasons. In the case of buffaloes, the number of HS infection cases were more in the rainy season (64.65 per cent), followed by winter (23.42 per cent) and summer (11.93 per cent) seasons. It showed that about 26 per cent cases occurred during the summer season in bovines in India (Singh *et al.*, 2014).

While stressful conditions contribute significantly to the outbreak of hemorrhagic septicemia in cattle and buffaloes, control of this disease through the control of stresses proved difficult (Moustafa*et al.*, 2015). Since the disease causes sudden death in animals, treatment of disease is in vain. Thus, vaccination appears to be the most common and cheap mean of controlling outbreaks of hemorrhagic septicemia. Vaccination against hemorrhagic septicemia was practiced as early as 1928. It is now widely practiced and has been adopted in most of the countries experiencing the disease (Ahmad *et al.*, 2014).

Research on control of hemorrhagic septicemia was carried out since 1960s (Thomas, 1968). Eventually, an effective hemorrhagic septicemia oil-adjuvant vaccine was produced in 1978 (Chandrasekaran and Yeap, 1978; Ahmad *et al.*, 2014). Eventhough, the disease could not be effectively contained in countries where hemorrhagic septicemia is endemic. HS is responsible for an economic loss of USD 784 million in India (Singh *et al.*, 2014) and the main reason for this was low the low vaccine coverage (Sahare and Salim,1991). Although many factors are found to influence the potency of hemorrhagic septicemia vaccine (Sarwar *et al.*, 2015), the main reasons for the failure are the ineffective vaccination regime and the poor vaccination coverage. The vaccination coverage should be at least 70 per cent (Neramitmansook, 1993). In spite of the recommended vaccination regime, most endemic countries practice vaccination once there is an outbreak. During an outbreak, vaccination is quickly done using broth bacterin vaccine to provide quick but low level coverage in the affected area (Chandrasekaran *etal.*, 1994). This is followed by ring vaccination in the surrounding areas using the oil adjuvant vaccine to provide a solid coverage that lasted for a year (De Alwis, 1993). Then the recommended yearly vaccination regime covering 70 per cent of the animal population was quickly forgotten until another outbreak. This stopgap measure in controlling hemorrhagic septicemia provides short-term solution but allows later outbreaks of hemorrhagic septicemia since frequency and timing of vaccination are important in controlling the disease. Vaccination coverage of less than 80 per cent may lead to the recurrence of disease outbreaks (Syamsudin, 1993).

Vaccines of Haemorrrhagic Septicaemia

Alum Precipitated Vaccine

It is the most widely used vaccine in Asia and Africa. It consists of a bacterin to which Potash alum has been added to give a final concentration of 1 per cent. The disadvantages of this type of vaccine are that it only provides reliable immunity for three to four months and shock reactions can also occur (Bain *et al.*, 1982). It is important to thoroughly mix the contents of the bottle before use because precipitation with alum causes the cells to settle at the bottom. If this precaution is not taken, animals receiving vaccine from the bottom of the bottle will receive a large dose of bacteria, and the chances of shock will be increased.

Aluminium Hydroxide Gel Vaccine

Aluminium hydroxide gel vaccine shares common properties with the alum precipitated vaccine. Aerated cultures are blended with aluminium hydroxide gel at a final concentration of 3 per cent. The protection by this vaccine will last for 4 months.

Oil Adjuvant Vaccine

Oil adjuvant vaccine (OAV) is considered as the most potent vaccine with 1 year protection (Muneer *et al.*, 2005). Commercial OAV against HS is formulated by emulsifying 50 per cent aqueous antigenic phase with 50 per cent oil phase by applying lanolin as emulsifier (OIE, 2008).While the oil adjuvant vaccine provides higher degree and longer duration of protection, the thick viscosity of the oil adjuvant vaccine is of major concern. It is difficult to be drawn into syringes and difficult to be administered in the field. Local swelling, sterile abscesses, and post-vaccination shock were reported in some animals following vaccination with the oil adjuvant vaccine (Joseph, 1979). Therefore, injectable oil adjuvant vaccine is less popular in the field although it has been shown to stimulate prominent and prolonged immunity. High viscosity and poor syringeability of the OAV is considered major constraint for its wide application. Many countries with endemic problem of haemorrhagic septicemia adopted the oil adjuvant vaccine for control of the disease (Ahmad *et al.*, 2014). The vaccination regime adopted for the oil adjuvant vaccine is yearly intramuscular vaccination that gives both a higher degree and a longer duration of immunity up to 1 year. For a maximum effect, vaccination should be given one month prior to the monsoon or rainy season (Bain *et al.*, 1982).

Efforts were made to reduce the viscosity of OAV by formulation of multiple emulsions (Verma and Jaiswal, 1997) and through replacing paraffin oil by Mercol 52 (Shah *et al.*, 1997), by changing the composition of oil and aquatic phase and also by co adjuvanting with aluminium hydroxide gel and saponin. The proportion of lanoline in the vaccine affects viscosity and injectability. The earliest oil adjuvant vaccines used around 8 per cent to 10 per cent lanoline. Experiments in Sri Lanka indicated that stable emulsions could be prepared using equal volumes of a light mineral oil and bacterin, with 4 per cent lanoline, thereby reducing the viscosity. Viscosity has also been reduced by incorporating Arlacei A as the emulsifying agent, together with Tween 80.

A thin, stable emulsion was formed using 40 parts of broth bacterin, 55 parts of mineral oil (Marcol 52), 4 parts of Arlacel A and 1 part of Tween 80 by weight (Neramitmansook, 1993). Reddy *et al.*, 1995 also reported long term immunity and protection in bovine against hemorrhagic septicemia for 1 year after primary immunization with Montanide adjuvanted vaccine. A new HS oil based vaccine has been developed by adding an adjuvant Montanide ISA-50V2 with the ratio of 1:1. The bacterial dry weight has been adjusted to 2 mg/ml which results in to reduction of dose per animal from 5 ml (alum precipitated) to 2 ml. The new vaccine has passed sterility, safety and potency tests as per OIE, 2017. This new product has low viscosity and single shot is expected to confer solid immunity against HS for one year.

Live Vaccine

The use of intranasal live attenuated vaccine allows self-vaccination and increases vaccination coverage among animal population (Saleem *et al.*, 2014). Intranasal administrations not only stimulate the mucosal immunity of the exposed host but also transmit the organism to the in-contact hosts and eventually stimulate their mucosal immunity. *Pasteurella multocida* mutants that are capable of providing heterologous protection against infection caused by virulent *P. multocida* has been patented (Garcia *et al.*, 2011). These mutants are defective in *fur*ompH and *fur*ompHgalE genes and are prepared as live vaccine to provide protection against infection by *Pasteurella multocida* including hemorrhagic septicemia.

Earlier, a *P. multocida* B:3,4 isolated from a fallow deer was used as a live vaccine was shown to provide partial protection in cattle and buffalo calves when given either subcutaneously or intradermally. This live vaccine was further improved by delivery *via* intranasal aerosol (Myint*et al.*, 1987). In cattle and buffaloes it has given 100 per cent protection against homologous serotypes. Also the serum of vaccinated animal provide passive cross protection in mice against infection with *Pasteurella multocida* serotypes E:2, F:3,4 and A:3,4. However, the vaccine has occasionally been associated with sporadic outbreaks of the disease (OIE, 2015) and occasional virulence to young animals, making it unacceptable.

Another attenuated live vaccine derived from a virulent *P. multocida* B:2 isolate (strain 85020) was developed by deletion of the *aroA* gene (Hodgson *et al.*, 2005). It was thought to be a suitable live vaccine candidate that, producing a high degree of protection against *P. multocida* B: 2 infection in mouse and calf models (Ataei *et al.*, 2009). However, upon experimental testing in cattle calves *via* the intramuscular and intranasal routes, it was observed that only the intramuscular vaccination route was able to provide protection against *P. multocida* B: 2 infection (Hodgson *et al.*, 2005). Thus, it was concluded that the *aroA* mutant strain is an effective live attenuated vaccine only if given *via* the intramuscular route.

Subunit Vaccines under Development

In case of *P. multocida*, the first attempt was non-toxic recombinant derivative of the *P. multocida* toxin (rPMT). Mohd Yasin *et al.* (2011)evaluated efficacy of an inactivated recombinant vaccine encoding a fimbrial protein of *P. multocida* B:2 against hemorrhagic septicemia in goats and concluded that inactivated recombinant vaccine significantly provides significant protection against high dose challenge and enhances the stimulation of the local and systemic immunities.

DNA Vaccines

DNA vaccine derived from *P. multocida* Toxin (PMT) gene was first developed. Outer membrane protein DNA vaccines such as OMP-DNA vaccines (pOmpH, pOmpA, pOmpHA), pOmpH, PlpEN and PlpEC, divalent combination of pcDNA-OmpH +pcDNA-OmpA, pOmpH +pOmpA, pcDNA-OmpH, pOmpH and pcDNA-OmpA, pOmpA and some others are the result of scientist attempts to achieve potent DNA vaccine against *P. multocida*. There are some advantages and disadvantages for this kind of vaccine preparation strategy like having no risk for infection, antigen presentation by both MHC class I and class II molecules, immune response focused on antigen of interest, ease of development and production, stability for storage and shipping, cost-effectiveness, not requiring peptide synthesis, expression.

Ghost Vaccine

P. multocida ghosts as new vaccine candidates and animals which received 1.15×10^8 ghosts and a challenge dose of up to 60 cfu produced 100 per cent protection. Thus ghosts of *P. multocida* can act as new vaccine candidates (Marchart*et al.*, 2003)

References

1. Ahmad TA, Rammah SS, Shewelta SA, Haroun M and El-Sayed LH, 2014. Development of immunization trials against Pasteurella multocida. Vaccine, 32: 909-917.
2. Ataei S, Burchmore R, Hodgson JC, Finucane A, Parton R *et al.*, 2009. Identification of immunogenic proteins associated with protection against hemorrhagic septicemia after vaccination of calves with a live-attenuated AroA derivative of Pasteurella multocida B: 2. Res Vet Sci, 87: 207-210.
3. Bain RVS, De Alwis MCL, Carter GR and Gupta BK, 1982. Haemorhagicsepticemia. FAO Animal Production and Health Paper No. 33. Food and Agriculture Organisation of the United Nations, Rome.
4. Chandrasekaran S and Yeap PC, 1978. Safety and potency testing of hemorrhagic septicemia oil adjuvant vaccine by mouse protection test. Kajian Vet, 10: 28-34.

5. Chandrasekaran S, Kennett L, Yeap PC, Muniandy N, Rani B *et al.*, 1994. Characterization of immune response and duration of protection in buffaloes immunized with hemorrhagic septicemia vaccines. Vet Microbiol, 41: 213-219.

6. De Alwis MCL, 1993. Pasteurellosis in production animals: a review. In: Pasteurellosis in Production Animals. ACIAR Proc No. 43: 11-12.

7. FAO (Food and Agriculture Organization) (1991) Proceedings of the Fourth International Workshop on Haemorrhagic Septicaemia, FAO-APHCA Publication.

8. Gajendragad MR and Uma S, 2012. Status of hemorrhagic septicemia in India. Indian Council of Agricultural Research (ICAR). Bengaluru, India. http: //nadres.res.in: 8080/Nadres_Uploads/UploadedFiles// VetEpiReports/ThirdSlot/Tech per cent 20HS per cent 20final.pdf.

9. Garcia JB, Saiz IB, Casas ML, Ocana MEG, Gallego MB *et al.*, 2011. Heterologous protection against Pasteurella multocida provided by Pasteurella multocida fur cells and the outer membrane proteins extracts thereof. US Patent, US 12/865, 287.

10. GoI (Government of India) (2007-2011) Annual Reports (2007-08 to 2011-12). Department of Animal Husbandry, Dairying and Fisheries, New Delhi.

11. Hodgson JC, Finucane A, Dagleish MP, Ataei S, Parton R *et al.*, 2005. Efficacy of vaccination of calves against hemorrhagic septicemia with a live aroA derivative of Pasteurella multocida B: 2 by two different routes of administration. Infect Immun, 73: 1475-1481.

12. Joseph PG, 1979. Hemorrhagic septicemia in Peninsular Malaysia. Kajian Vet, 11: 65-79.

13. Marchart J, Dropmann G, Lechleitner S, Schlapp T, Wanner G, Szostak MP, *et al.*, 2003.P.multocida- and Pasteurella haemolytica ghosts: new vaccine candidates. Vaccine, 21(25-26): 3988-3997.

14. Mohd Yasin IS, MohdYusoff S, Mohd ZS, Abd Wahid Mohd E. 2011.Efficacy of an inactivated recombinant vaccine encoding a fimbrial protein of P. multocida B: 2 against hemorrhagic septicemia in goats. Trop Anim Health Prod., 43(1): 179-187.

15. Moustafa AM, Ali SN, Bennett MD, Hyndyman TH, Robertson ID *et al.*, 2015. A case-control study of heamorrhagicsepticemia in buffaloes and cattle in Karachi, Pakistan, in 2012. Trans Emerg Dis, doi: 10.1111/tbed.12393.

16. Muneer R, Hussain M, Zahoor A. 2005.Efficacy of oil based haemorrhagic septicaemia vaccine: A Field Trial. Int J Agr Biol. 7: 571–573.

17. Myint A, Carter GR and Jones TO, 1987. Prevention of experimental hemorrhagic septicemia with a live vaccine. Vet Rec, 120: 500-501.

18. Neramitmansook P, 1993. Country Report: Thailand. In: Pasteurellosis in Production Animals. ACIAR Proc No. 43: 234-237.No. 1991/13.

19. OIE 2015. Manual of standards for diagnostic tests and vaccines for terrestrial animals. http: //www.oie.int/international-standardsetting/ terrestrial-manual/access-online/

20. OIE. 2008. Haemorrhagic septicaemia OIE Terrestrial Mannual, chapter 2.4.12. Pp 739–51.

21. Reddy, G.S., Rao KA and Srinivasan, VA 1995. Immunity conferred by oil adjuvant haemorrhagic septicaemia vaccine. Ind J Anim Sci., 66(7): 703–704.

22. Saharee AA and Salim NB, 1991. The epidemiology of haemorrahgicsepticemia in cattle and buffaloes in Malaysia. Proc 4th Intl Workshop Hemorrhagic septicemia, Kandy, Sri Lanka, 109-112.

23. Saleem L, Munir R, Ferrari G, Afral M and Chaudhry FR, 2014. Efficacy and cross-protection of live intranasal aerosol hemorrhagic septicemia vaccine in buffalo calves. Int J Curr Microbial Appl Sci, 3: 300-307.

24. Sarwar N, Muhammad K, Rabbani M, Rana MY, Sarwar M *et al.*, 2015. Factors affecting the potency of hemorrhagic septicemia vaccines. Int J Agric Biol, 17: 387-390.

25. Shah NH, Shah NH and Graaf FKD. 1997. Protection against haemorrhagic septicaemia induced by vaccination of buffalo calves with an improved oil adjuvant vaccine. FEMS Microbiology letters **115**: 203–07.

26. Singh B, Prasad S, Verma MR and Sinha DK, 2014. Estimation of economic losses due to hemorrhagic septicemia in cattle and buffaloes in India. Agric Econ Res Rev, 27: 217-279.

27. Singh B.Pra**sad** S. Verma MR and Sinha DK 2014.Estimation of economic losses due to haemorrhagic septicaemia in cattle and buffaloes in India. Agricultural Economics Research review 27 (2) 271-279.

28. Tasneem, K., Butt, IA., Raza I. and Gill, Z.J. 2003. Biochemical serological and immunological properties of Pasteurella multocida strains isolated from natural out breaks of Haemorrhagic septicaemia. Pak J Vet Res., 1(1): 1-4.

29. Thomas J, 1968. Studies on hemorrhagic septicemia oil adjuvant vaccine. I. Method of production of vaccine. Kajian Vet, 1: 152-158.

30. Verma R and Jaiswal T N. 1997. Protection, humoral and cell mediated immune response in calves immunized with multiple emulsion HS vaccine. Vaccine **15**: 1254–60.

Chapter 5

Diagnosis of Aquatic Animal Diseases: Challenges and Future Dimensions

K.J. Reshma* and T.G. Sumithra

Scientist, Marine Biotechnology Division,
ICAR-Central Marine Fisheries Research Institute, Kochi – 682 018, Kerala,
**e-mail: reshjanardhan@gmail.com*

Global food fish consumption augmented at an average annual rate of 3.1 per cent from 1961 to 2017, which were recorded to be higher than that of all other animal protein foods (meat, dairy, milk, *etc.*), which increased by 2.1 per cent per year. Among which, aquaculture accounted for 46 per cent of the total global fish production and contributes 52 per cent of fish for human consumption. Interestingly, aquaculture scored as the fastest growing food-producing sector in the world, with an average annual growth rate of 7.5 per cent since 1970, and 2.6 per cent for terrestrial farmed meat production systems over the same period (FAO, 2020). Concurrently, the disease situation in aquaculture is taking a new horizon due to the current period of accelerated change in the international trading environment affected by globalization, increasing aquaculture production, microbial adaptation and climate change (Woo *et al.*, 2020). As of now, disease problem has already turned out to be a primary constraint to the aquaculture, in tandem with advancement in technological interventions, impeding both economic and social development in many countries (Tacon, 2020).

Diseases in Aquatic Organisms: Are we Burning our Bridges?

Similar to any food production system, growing aquaculture production will come with environmental costs. Local and regional environmental factors significantly determine ecosystem impacts of production, such as those related to accumulation of effluent and sensitivity of wild species (Aguilar-Manjarrez *et al.*, 2017). Most of the researchers suggest that intensive farming practices affect the environment and may drive disease emergence in both terrestrial and aquatic animals. Alternatively, some researches pointed out that all new diseases that arise in farmed aquatic animals have their origins in wild sympatric populations. However, a considerably superior degree of connectivity and exchange of pathogens occurs in the aquatic environment as compared to the terrestrial one, and conditions in many aquaculture systems, like high density of monocultured, genetically homogenous hosts, favour virulent strains (Kurath and Winton, 2011). It is further anticipated that global climate change will trigger an increased occurrence of disease events, especially when anthropogenic movements of aquatic animals such as relocation of farming species are taken into account. This will augment the host factors related to physiological stress and associated decrease in immunocompetence may allow existing components of the host pathobiome to become more virulent, or facilitate infections with opportunistic pathogens that are present in the local environment. The scenario is likely to be worsen, as the rate of emergence of new diseases in aquaculture systems has amplified in recent decades (Feist *et al.*, 2019) but the impacts of disease in wild populations yet to be unravelled, with their associated socioeconomic consequences (Miller *et al.*, 2014). In most of the poorly managed scenarios, rolling back environmental and animal health regulation results in disease outbreaks and biodiversity loss (Gephart *et al.*, 2021) and ineffective policies and lax regulations have resulted in disease outbreak and collapse of the aquaculture production (Costello *et al.*, 2020). However, understanding the environmental boundaries, improved knowledge about disease identification, prevention and treatment are the possible remedies to tackle disease problems in aquaculture (Gephart *et al.*, 2021). Timely diagnosis of the problem will help to choose suitable management practices within the farm and determine if future outbreaks could be prevented by changes in procedure or design.

Diagnosis of Aquatic Animal Diseases: Fishing for Solutions

Diagnosing aquatic animals by the clinical sign is almost difficult because these animals live in aquatic medium and move fast that make them impossible to physically visualize and closely inspect for any clinical abnormalities (Assefa and Abunna, 2018). So periodical sampling is required for them to screen the presence of pathogens, or accompanying diseases, so that scientifically-based treatments can be applied before significant losses occur (Austin and Austin, 2016). Water quality conditions are also given equal weightage in aquatic animal disease investigations, as incidence and severity of infectious diseases in these animals are very often dependant on the quality of the aquatic environment (Mishra, 2010). With a few modifications, most of the diagnostic procedures for handling of the aquatic animal patient, resembles to procedures performed in most veterinary hospitals and clinics of other companion species (Reavill, 2006). However, the predictive value of these diagnostic methods is improving by adoption of advanced technologies from related fields with ever-increasing number of publications in aquatic animal medicine.

Triage and Patient Assessment

Due to the existing limitations of onsite evaluation of aquatic animal diseases for the clinical expert, in many cases, the abnormal, diseased or moribund live animal has to be transported to the nearest aquatic animal clinic with adequate precautions. The sampled fish should be sick and demonstrating symptoms, but alive on transportation. In most cases, it is suggested that a water sample should be submitted for basic water chemistry testing along fish sample submission, so that water quality can be ruled out as the cause or a causal factor to fish mortalities/disease (Reavill, 2006). All the samples should be maintained at lower temperature and time for transportation must be reduced to 4 hours. An assessment of clinical signs and case history (as much as information about the circumstances and appearance of the disease condition) may be made during the initial personal/phone consultation, where the owner is being instructed on how to bring in the sample. This information combined with laboratory identification of the infectious agent gives the investigator the most complete picture of the problem. If possible, the clinical expert can go for an on-site examination; where, every effort is made to explore even the most unlikely possible causes. In cases where mortalities are initiated by external parasites and bacteria, examining material on site is the best way to identify the problem.

Clinical Techniques: Routine Laboratory Procedures

Once the samples are collected in lab, the animals are observed for clinical signs typical of a particular disease, in such cases, the agent could be isolated using specific methods and diagnosis can be accomplished relatively quickly. Water quality and other environmental parameters need to be assessed concurrently to assess their contributions to disease outbreaks. Nutritional deficiencies and even genetic aberrations are also essential to be considered for understanding the apparent health of organism. Further, pattern of mortality has to be closely monitored as mortalities that seem to have a uniform or random distribution should be examined immediately. Analysis of the samples protocols for fish disease diagnosis varies from laboratory to laboratory. However, a general protocol for aquatic animal disease diagnosis is depicted in Figure 5.1.

As an initial step, the sampled aquatic animal was observed for any physical or behavioural abnormalities in the laboratory. The physical abnormalities include, discolorations, erosions, deformities, damaged shells, dropsy, dysecdysis and haemorrhages. On the other hand, sick animal always tries to isolate themselves from the healthy cohort and show some behavioural abnormalities like erratic movements, gulping, flashing and rubbing against the hard surfaces. The major necropsy techniques performed in laboratory includes blood, tissue collection and gill/skin biopsy. Collection of the blood is usually done non lethally through the caudal veins for fishes. The needle is introduced at the ventral midline of the caudal peduncle, caudal to the anal fin. Alternatively, hemolymph is being collected from invertebrate heart or walking leg sinus. Faecal sample collections are usually carried out for parasitic investigations (Noga, 2010).

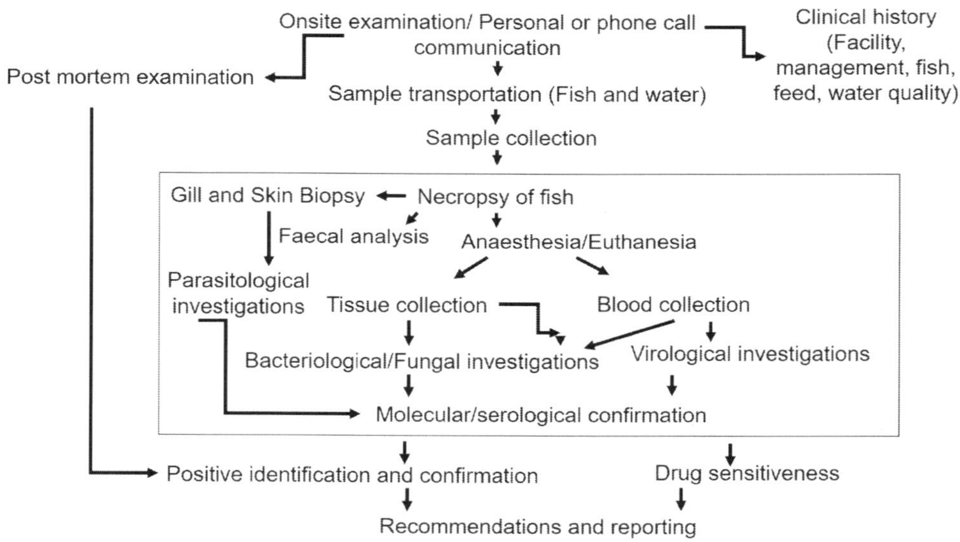

Figure 5.1

Diagnosing Aquatic Animal Disease: Off the Mark or Hit the Mark? - An Avenue for Development

The diagnosis of aquatic animal is comparatively a difficult task, as the clinical experts never get chance for a direct examination of the patient. Moreover, aquatic animals cannot be observed close enough like we do in terrestrial animal and the environment can facilitate disease transmission quickly. The diagnosis of aquatic animals is further get complicated, as there is difficulty in catching these animals without imposing stress during sampling (Peeler and Taylor, 2011). The transportation itself is a stress to animals as they are prone to low levels of oxygen conditions and high levels of ammonia wastes during transportation. Further, the other important challenge in diagnosing aquatic animal's disease is that the unit of interest is not a single animal, rather the whole population need to be investigated and diagnosed, especially when the prevalence of disease is low. Further, samples should be collected not only from animals but also from residing water; which makes aquatic animal diagnosis complicated and challenging (Ninawe *et al.*, 2017). There may be problems in interpreting clinical signs of aquatic animals, as acute and chronic phases of most of the diseases have different signs. Therefore, a clinical expert has to reach the diagnosis with a partial set of signs. In most cases, a disease is identified, when the first clinical signs appear and will exhibit different signs than initial phase when mortalities are already occurring. In many cases, by the time the fish exhibit signs of a problem, there is already a secondary agent involved. In many occasions, the signs observed by the farmer may indicate a co-infection. The diagnosis of aquatic animal disease is complicated, as there is a considerable biasedness in signs chosen by a farmer to describe a condition, as this is influenced by his knowledge and experience (Zeldis and Prescott, 2000). Further, clinicians should be vigilant while interpreting the haematological characters of aquatic animals, as environmental parameters and nutritional status have been shown to significantly affect these values (Divincenzi *et al.*, 2013) Further, fish clinicians should be very much aware of the lack of evidence supporting direct extrapolation from domestic mammal clinical pathology results in fish (Vergneau-Grosset and Larrat, 2017). Microbiological tests, are used extensively in aquatic animals but care should be taken for microbiological investigation of invertebrates who display an open vascular system with constant and direct contact with the environment. Further, many of healthy invertebrates normally display positive blood cultures (Tao *et al.*, 2014). Moreover, the temperature of incubation as well as media of isolation of microbiological analysis should be close to the animal's residing environment to avoid *in vitro* selection of different germs. In culture dependant bacteriological methods, it has been unclear what proportion of bacterial cells actually produces colonies on solid media. In such situations, acquisition of cultures does not necessarily mean that the actual pathogen has been recovered, but could reflect recovery of secondary invaders of already diseased tissue or even contaminants from the aquatic environment. Usually, Bacteriological examination on a single occasion within a disease cycle of aquatic animals would not identify possible microbial succession, whereby an organism recruits an infection but subsequently become out competed by others opportunistic pathogens. Moreover, the time needed to get and identify cultures is more in microbiological identifications, when the real thrust is relied on controlling the disease. (Austin, 2019).

Within the boundaries of challenges, disease diagnosis of aquatic animals is slowly emerging with advancement of technologies and new validated protocols. As rapid detection of disease clinically or sub-clinically is of utmost important to prevent crop loss in culture systems, FAO/NACA (2000) has recommended three levels of diagnostics, based on the scale of expertise and infrastructure availability. Of the three levels level III is considered more advanced and recognised to produce quicker results. The evolution of routine diagnosis from direct microscopy, histology to electron microscopy increased precision in diagnostic techniques. Currently, most of the aquatic animal disease clinics are using PCR and associated variations for rapid diagnosis. Post twentieth century, with an embracing developments in molecular biology, culture-independent techniques took a new turn in bacteriological identifications with sequencing of the *16S rRNA* gene, which does not need intact, viable bacterial cells (Austin, 2019). The approaches had more speed and accuracy with high level of specificity and the bacteria may be studied and identified regardless of whether or not they may be grown in the laboratory (Austin, 2017). In addition, accurate identification of cultured colonies may be puzzling and development of molecular diagnostic techniques, including matrix-assisted desorption ionization time-of-flight mass spectrometry is a latest advance in fish disease diagnosis (Olate *et al.*, 2016). Recently thrust is also given to apply artificial intelligence (AI) for the rapid and accurate diagnosis of diseases in fishes, but considerable effort has yet to put upon increasing the feasibility of AI by supplying an adequate database of fish diseases (Lopes *et al.*, 2011).

References

1. Adams, A., 2009. Advances in disease diagnosis, vaccine development and other emerging methods to control pathogens in aquaculture. *New Technologies in Aquaculture: Improving Production Efficiency, Quality and Environmental Management*, pp.197-214.
2. Aguilar-Manjarrez, J., Soto, D. and Brummett, R., 2017. *Aquaculture zoning, site selection and area management under the ecosystem approach to aquaculture. A handbook*. Food and Agriculture Organization of the United Nations (FAO).
3. Amos, K.H., 1985. *Procedures for the detection and identification of certain fish pathogens*. Fish Health Section, American Fisheries Society.
4. Assefa, A. and Abunna, F., 2018. Maintenance of fish health in aquaculture: review of epidemiological approaches for prevention and control of infectious disease of fish. *Veterinary medicine international, 2018*.
5. Austin, B. and Austin, D.A., 2016. *Bacterial fish pathogens: disease of farmed and wild fish*. Springer.
6. Austin, B., 2017. Antibiotics and disinfectants. *in Diagnosis and Control of Diseases of Fish and Shellfish*, pp.263-278.
7. Austin, B., 2019. Methods for the diagnosis of bacterial fish diseases. Marine Life Science and Technology, 1(1), pp.41-49.
8. Costello, C., Cao, L., Gelcich, S., Cisneros-Mata, M.Á., Free, C.M., Froehlich, H.E., Golden, C.D., Ishimura, G., Maier, J., Macadam-Somer, I. and Mangin, T., 2020. The future of food from the sea. *Nature, 588*(7836), pp.95-100.
9. DiVincenti, L., Priest, H., Walker, K.J., Wyatt, J.D. and Dittman, D., 2013. Comparison of select hematology and serum chemistry analtyes between

wild-caught and aquarium-housed lake sturgeon (Acipenser fulvescens). *Journal of Zoo and Wildlife Medicine*, *44*(4), pp.957-964.

10. FAO, IFAD, UNICEF, WFP and WHO. 2020. *The State of Food Security and Nutrition in the World 2020. Transforming food systems for affordable healthy diets*. Rome, FAO.
11. Feist, S.W., Thrush, M.A., Dunn, P., Bateman, K. and Peeler, E.J., 2019. The aquatic animal pandemic crisis. *Revue scientifique et technique (International Office of Epizootics)*, *38*(2), pp.437-457.
12. Gephart, J.A., Golden, C.D., Asche, F., Belton, B., Brugere, C., Froehlich, H.E., Fry, J.P., Halpern, B.S., Hicks, C.C., Jones, R.C. and Klinger, D.H., 2020. Scenarios for global aquaculture and its role in human nutrition. *Reviews in Fisheries Science and Aquaculture*, *29*(1), pp.122-138.
13. Kurath, G. and Winton, J., 2011. Complex dynamics at the interface between wild and domestic viruses of finfish. *Current Opinion in Virology*, *1*(1), pp.73-80.
14. Lopes, J.N.S., Gonçalves, A.N.A., Fujimoto, R.Y. and Carvalho, J.C.C., 2011. Diagnosis of fish diseases using artificial neural networks. International Journal of Computer Science Issues (IJCSI), 8(6), p.68.
15. Miller, K.M., Teffer, A., Tucker, S., Li, S., Schulze, A.D., Trudel, M., Juanes, F., Tabata, A., Kaukinen, K.H., Ginther, N.G. and Ming, T.J., 2014. Infectious disease, shifting climates, and opportunistic predators: cumulative factors potentially impacting wild salmon declines. Evolutionary Applications, 7(7), pp.812-855.
16. Mishra, S., 2010. Fish disease management in integrated farming system.
17. Ninawe, A.S., Hameed, A.S. and Selvin, J., 2017. Advancements in diagnosis and control measures of viral pathogens in aquaculture: an Indian perspective. *Aquaculture International*, *25*(1), pp.251-264.
18. Noga, E.J., 2010. *Fish disease: diagnosis and treatment*. John Wiley and Sons.
19 Olate, V.R., Nachtigall, F.M., Santos, L.S., Soto, A., Araya, M., Oyanedel, S., Díaz, V., Marchant, V. and RiosMomberg, M., 2016. Fast detection of Piscirickettsia salmonis in Salmo salar serum through MALDITOFMS profiling. *Journal of Mass Spectrometry*, *51*(3), pp.200-206.
20. Peeler, E.J. and Taylor, N.G., 2011. The application of epidemiology in aquatic animal health-opportunities and challenges. *Veterinary Research*, *42*(1), pp.1-15.
21. Reavill, D.R., 2006. Common diagnostic and clinical techniques for fish. *Veterinary Clinics: Exotic Animal Practice*, *9*(2), pp.223-235.
22. Tacon, A.G., Lemos, D. and Metian, M., 2020. Fish for health: improved nutritional quality of cultured fish for human consumption. *Reviews in Fisheries Science and Aquaculture*, *28*(4), pp.449-458.

23. Tao, Z., Bullard, S.A. and Arias, C.R., 2014. Diversity of bacteria cultured from the blood of lesser electric rays caught in the northern Gulf of Mexico. *Journal of aquatic animal health, 26*(4), pp.225-232.

27. Vergneau-Grosset, C. and Larrat, S., 2017. Evidence-based advances in aquatic animal medicine. Veterinary Clinics: Exotic Animal Practice, 20(3), pp.839-856.

25. Walker, P. and Subasinghe, R.P. eds., 2000. *DNA-based molecular diagnostic techniques: research needs for standardization and validation of the detection of aquatic animal pathogens and diseases* (No. 395). Food and Agriculture Org.

26. Woo, P.T., Leong, J.A. and Buchmann, K. eds., 2020. *Climate Change and Infectious Fish Diseases*. CABI.

27. Zeldis, D. and Prescott, S., 2000. Fish disease diagnosis program—problems and some solutions. *Aquacultural Engineering, 23*(1-3), pp.3-11.

Chapter 6

Application of Immunohistochemistry in Disease Diagnosis

Deepika Lather, Babu Lal Jangir, Vikas Nehra, Gauri Chandratre A. and Gulshan Narang

Department of Veterinary Pathology, College of Veterinary Sciences, Lala Lajpat Rai University of Veterinary and Animal Sciences, Hisar – 125 004, Haryana

Immunohistochemistry (IHC) is one of the most powerful techniques in diagnostic pathology which is widely used in diagnosis of diseases associated with autoantibody deposition, infectious disease microorganisms and for the identification of tumours. The technique combines the three basic disciplines *i.e.* immunology, histology and chemistry. It is a quick, sensitive, specific, and relatively inexpensive technique. It employs use of specific antibodies to viral, bacterial, fungal, parasitic and prion antigens in the detection and identification of the causative agents. Detecting animal diseases by this technique will allow an early diagnosis of such diseases and reduce the effect of infections in herds. It is also convenient with respect to submission of specimens. It can be performed on formalin fixed, paraffin-embedded tissue, reducing the risk of exposure to serious infectious diseases; allows the identification of infectious agents by monoclonal antibodies and some polyclonal antibodies even before morphologic changes occur in the tissues. But the only limitation in the technique is the availability of specific antibodies and the masking of antigenic epitopes. It is pertinent to mention that both monoclonal and polyclonal antibodies must be tested and standardized for possible cross-reactivity with other organisms before recommended use. The applications of IHC have been recently increased due to elucidation of role of various molecules in pathogenesis, diagnosis and treatment of diseases.

Application in Infectious Diseases

In diagnosis and pathogenesis of infectious diseases IHC plays a very important role. The detection of the organism in the lesion helps to confirm that it is involved in the pathogenesis of the disease. Detection of microorganisms in processed tissue specimens also facilitates retrospective studies. Most diagnostic laboratories store, paraffin-embedded blocks of the samples of all submitted specimens, and most antigens are stable indefinitely in tissues within paraffin blocks thus permitting long term retrospective studies for study of the incidence and prevalence of diseases.

Viral infections are usually characterized by production of characteristic cytomorphological changes such as intracellular inclusions, which allow presumptive diagnosis. But these inclusions are sometimes inconspicuous or very few in number and hematoxylin-eosin stained sections could not demonstrate them. By immunoperoxidase localization, viral core antigens can be demonstrated within the nuclei or the cytoplasm of infected cells. Also for differentiation between the diseases producing similar types of inclusions IHC become important for confirmation. Another advantage is that it allows the detection of the antigens early in the course of the disease when cytopathic changes have not yet developed. The number of immunohistochemical tests for disease surveillance for viral pathogens has increased progressively over the last few years *e.g.* Rabies, Infectious bovine rhinotracheitis, Bovine viral diarrhoea, rotavirus, coronavirus, adenoviruses, bovine respiratory syncytial virus, equine and bovine herpes virus, transmissible gastroenteritis, porcine circovirus 2, porcine reproductive and respiratory virus and proliferative enteritis of pigs.

Heavy bacterial infections with large number of organisms are easily detected by routine hematoxylin and eosin or other histochemical stains. But for the detection of scant numbers of organisms, IHC has proved to be superior to conventional methods. For bacterial pathogens, antigen retrieval is generally not required in fixed tissues. But many of the antibodies cross-react with each other so interpretation of the results can be complicated. Moreover, antibodies may react with only portions of the bacteria, and they may label remnants of bacteria or spirochetes when viable organisms are no longer present. This method is advantageous in cases of bacterial infections in which the likelihood of culture is reduced due to systemic treatment of the animal with antibiotics.IHC has been found to be useful in the diagnosis of Campylobacteriosis, Brucellosis, Pasteurellosis, Salmonellosis, leptospirosis *etc.*

Fungal infections mainly caused by *Aspergillus* spp. or zygomycetes, are considered to be opportunistic agents that are usually secondary to other conditions and disrupt the immune status of the host. Systemic mycoses often involve several organs, *e.g.* the lungs, mammary gland or fore-stomachs, resulting in a multifocal pyogranulomatous inflammatory disease. Mycoses are difficult to diagnose clinically and fungi are relatively difficult to culture as they often require specialised growth media and conditions. Fungal organisms are generally readily identified by hematoxylin and eosin staining alone or in combination with histochemical stains [periodic acid–Schiff (PAS)], and Gomori's methenamine silver (GMS). However, these stains cannot distinguish morphologically similar fungi. IHC has been used for identification of various fungal elements in paraffin-embedded, formalin fixed tissue. Commercially available monoclonal antibodies have been used successfully to demonstrate the presence of fungal hyphae within necrotic lesions. This method has the advantage of providing rapid and specific identification of several fungi and allowing pathologists to be able to identify unusual filamentous hyphal and yeast infections and accurately distinguish them from confounding artifacts. But prevalence of common antigens among pathogenic fungi frequently results in cross-reactivity with polyclonal antibodies and even with some monoclonal antibodies.

Among the parasitic infections the role of IHC has been limited to cases in which the morphology of the parasite is distorted by tissue necrosis or autolysis or infections with unusual presentations.

Technical Considerations

The basic principle of IHC is the detection of antigens (Ag) within tissue sections by means of specific antibodies (Abs). Once binding of specific antigen-antibody occurs, it specifically visualizes the antigen distribution in tissue sections by a coloured histochemical reaction under light microscope. There are basically two methods of IHC staining- Direct and Indirect. Direct staining methods are those in which the primary antibody (antibody specific for the antigen of interest) has been directly labelled with the enzyme or fluorescent dye, whereas in indirect techniques the primary antibody is unlabelled. In Indirect immunostains, the binding of the primary antibody to the tissues is detected by a second antibody to immunoglobulin which has been labelled with the enzyme or fluorescent dye. Indirect methods are a means of increasing the sensitivity of the staining method so that smaller amounts of antigen may be detected. The basic IHC protocol includes many critical steps such as proper collection and handling of specimen, fixation, tissue processing, antigen retrieval, antibody selection, preparation of reagents, incubation conditions, proper washing in between different steps, counter staining and finally the interpretation of results.

Basic Steps of IHC

The following are the basic steps involved in the IHC:

1. Fixation and processing of tissues
2. Antigen retrieval
3. Quenching of endogenous peroxidase activity
4. Immunohistochemical staining procedure

1. Fixation and Processing of Tissues

IHC can be performed on acetone fixed frozen tissue (cryostat sections) or FFPE (formalin fixed paraffin embedded) sections using appropriate antibodies. The proper fixation of tissue is the key step in IHC. In histopathology, the coagulative and non-coagulative (cross-linking) types of fixatives are used. The coagulative fixatives exert their preservative action by denaturing the proteins and non-coagulative fixatives do so by cross-linking the adjacent proteins with covalent bonds. Formaldehyde (aldehyde fixatives), methanol and ethanol (protein-denaturing agents) and acetone are the most commonly used fixatives for the immunochemical staining techniques. 10 per cent neutral buffered formalin and 4 per cent paraformaldehyde solution are routinely used. For most of the tissues, fixation for 24 hours at room temperature (RT) with appropriate tissue to fixative ratio *i.e.* 1:10-1:20 is recommended. In IHC, staining quality and intensity is mainly affected by the duration of fixation. Antigen retrieval can unmask the antigens; however, if the tissue has been fixed over a longer duration then there may be no signal even after antigen retrieval. After proper fixation, tissue samples are washed in running tap water to remove formalin and then processed in ascending grades of ethanol for dehydration, cleared in benzene, infiltrated and embedded in paraffin.

The 3-4 µm thick tissue sections are cut with microtome and mounted on 3-Aminopropyl-triethoxy-silane (APES) coated slides. Thereafter, these slides are placed in oven at 60°C for 1 hour for removal of paraffin for proper spreading of the sections. The higher background signals may be noticed in case of thick tissue sections as occurs in frozen sections. The incomplete removal of the paraffin can lead to improper staining of sections.

2. Antigen Retrieval

During fixation, masking of antigenic sites occurs due to cross linking of proteins by formation of methylene bridges. Heat induced epitope retrieval (HIER) and enzymatic methods are the most commonly used methods for antigen retrieval. Some antigens prefer heat mediated antigen retrieval to enzymatic and vice versa. The best retrieval condition for each Ag-Ab pair needs to be determined by comparing the results of staining obtained by employing different methods of antigen retrieval.

The HIER is generally performed by using a microwave oven, pressure cooker or vegetable steamer. If the antigen retrieval method is not stated on antibody datasheet, in such cases the optimal method for each antigen must be experimentally determined. The antigen retrieval with sodium citrate (0.01M, pH 6.0) or Tris/EDTA (pH 9.0) buffers are widely used. Antigen retrieval is performed by subjecting the tissue sections to microwave irradiation at 800W (as per standardization) in a coplin jar containing any of the above mentioned buffers (as per standardization) for 30 minutes (15 cycles of 2 minutes each). The slides are then cooled to RT and subsequently washed with phosphate buffer saline (PBS, pH 7.2) thrice (for 5 minutes each).

For enzymatic antigen retrieval, many enzymes such as trypsin, proteinase K, pronase, ficin and pepsin have been used for this purpose. Many times, the enzymatic choice is mentioned on the datasheet for the antibody. If not, then trypsin is used as it has been shown to be useful for a wide range of antigens. The enzyme solution can be applied by directly pipetting the solution over the tissue section or placing the tissue slides into a container filled with enzyme solution. Tissue sections are incubated with either trypsin or proteinase K for 10-20 minutes at 37°C and the reaction is then terminated by adding PBS. However, care must be taken as excessive exposure to these techniques can damage tissue sections degrading their morphological appearance and/or resulting in excessive non-specific binding of the primary antibody.

3. Quenching of Endogenous Peroxidase Activity

Slides are placed in coplin jar filled with 0.3-3 per cent hydrogen peroxide (H_2O_2) solution (in absolute methanol) for 15-30 minutes to block the endogenous peroxidase activity. Then, slides are thoroughly rinsed with PBS (pH 7.2) thrice for 5 minutes each. Some antigens such as CD4 can be destroyed by 3 per cent H_2O_2. In such cases, 0.5 per cent H_2O_2 concentration is used.

4. Immunohistochemical Staining

Availability of Specific Antibodies and Optimization of Antibody Dilutions

All diagnostic IHC requires the availability of a validated primary antibody of known specificity to the pathogen or disease-associated marker of interest. Primary antibodies used can be raised or validated in the laboratories or sometimes human and mouse antigens that cross-react with animal proteins can also be used. Datasheets often list known cross reactivities of the antibody. However, all primary antibodies require validation for specificity and sensitivity of the test. Validation involves using both positive and negative control preparations.

Optimization determines the optimal conditions required by the primary antibody, such as the required dilution of the primary antibody, incubation time and temperature and the consistency of the blocking solution to prevent non-specific binding. It is carried out by making serial dilution such as undiluted, 1:10, 1:20, 1:50, 1:100, 1:200, 1:500, 1:1000 *etc.* in 1 per cent bovine serum albumin (BSA) prepared in PBS and subjecting it to standard procedure for IHC. The antibody dilution which gives the best result is further used for the immunohistochemical staining. All incubation steps should be performed in humidified chamber to prevent the drying of tissues. The non-specific binding occurs if tissue sections are dried at any stage and ultimately gives high background staining.

Visualisation Systems

The various enzyme substrate combinations can be used in IHC for visualization of the results. One of the most common enzymes used in Immunohistochemical staining is peroxidase. Horseradish peroxidase (HRP) can be used in combination with 3,3′diaminobenzidine (DAB) or 3-amino-9-ethyl carbazole (AEC) as substrate which gives brown and brick red colour, respectively.

Immunohistochemistry Protocol

The brief protocol for the immunohistochemical detection of the antigen of infectious agents or antigens of either diagnostic or prognostic significance in tumours in formalin fixed paraffin embedded tissue sections is described below for reference:

- ✰ Deparaffinization of tissue sections on APES coated slides by immersing in xylene and then rehydration in the descending grades of ethanol (100 per cent, 95 per cent, 90 per cent, 80 per cent, 70 per cent and 50 per cent)
- ✰ Antigen unmasking by subjecting the tissue sections to microwave irradiation at 800W (as per standardization) in a coplin jar filled with citrate buffer (0.01M, pH 6.0) for 30 minutes (15 cycles of 2 minutes each)
- ✰ Then cooling of slides to RT and rinsing thrice with PBS (pH 7.2) for 5 minutes each
- ✰ Thereafter, quenching of endogenous peroxidase activity by dipping sections in 3 per cent H_2O_2 solution prepared in absolute methanol for 30 minutes
- ✰ Washing of slides thrice with PBS (pH 7.2) for 5 minutes each
- ✰ Excess fluid on sections is drained off and careful wiping around sections
- ✰ Encircling of sections with a hydrophobic pen
- ✰ Blocking of the non-specific sites by incubating sections with 5 per cent normal goat serum prepared in 1 per cent BSA in PBS (pH 7.2) for 45-60 minutes at 37°C in a humidified chamber (An ideal agent for blocking of non-specific sites is 5-10 per cent normal serum of same species in which secondary antibody is raised). Similarly, the blocking of endogenous alkaline phosphatase (AP) should be done with levamisole at a concentration of 10 mMas it is prevalent in the frozen tissues.
- ✰ Then washing of slides thrice with PBS (pH 7.2) for 5 minutes each (Sufficient washing after the blocking step is critical to remove excess protein that may prevent detection of the target antigen)

- ☆ Excess fluid on sections is drained off and careful wiping around sections
- ☆ Cover sections with the primary antibody diluted in 1 per cent BSA prepared in PBS (pH 7.2) at optimum dilution (as standardized by serial dilution as mentioned earlier in text)
- ☆ Tilting of the sections for 2-3 times in clockwise and anticlockwise direction for uniform spreading of antibody
- ☆ Incubation for overnight at 4°C or 37°C for 1-2 hours in a humidified chamber
- ☆ After overnight incubation, slides kept at RT for 15 minutes and then washing thrice with PBS for 5 minutes each
- ☆ Excess fluid on sections is drained off and careful wiping around sections done
- ☆ Cover sections with conjugated secondary antibody diluted in 1 per cent BSA prepared in PBS (pH 7.2) at optimum dilution (as per data sheet of manufacturer or standardized by serial dilution)
- ☆ Tilting of the sections for 2-3 times in clockwise and anticlockwise direction for uniform spreading of antibody
- ☆ Incubation for overnight at 37°C for 30-45 minutes in a humidified chamber
- ☆ Wash the slides thrice with PBS for 5 minutes each
- ☆ Excess fluid on sections is drained off and careful wiping around sections
- ☆ Cover sections with sufficient substrate solution (3-Amino-9-ethyl-carbazole) prepared as per manufacturer's instructions
- ☆ Terminate the reaction before appearance of generalized background staining in negative control by rinsing with distilled water
- ☆ Counter stain sections lightly with Gill's haematoxylin
- ☆ Rinse with distilled water and then mount in aqueous mounting medium (CC mount)

Note: Always put known positive and negative control. For negative control, cover the duplicate section with diluent only i.e., without primary antibody or use known tissue section which is devoid of particular antigen of the known infectious antigen.

Interpretation

Brick red or reddish-brown staining is considered as positive immunostaining.

Evaluation and Scoring of Immunostaining

The IHC data should be expressed in the numerical values for statistical analysis. The results should be assessed as either positive or negative. Samples that display no staining or staining in less than a certain percentage of positive cells are considered as negative while the samples that show unequivocal staining in at least a certain percentage of cells are considered positive. During the evaluation of results of IHC, the researchers generally assess the relative percentage of immunopositive cells in relation to total number of target cells. This method has certain limitation that it does not depict staining intensity. Therefore, combinative semi-quantitative scoring method is preferred as it yields the immunoscores that incorporate both quantitative and qualitative assessment. In addition to the quantitative data by assessing the relative percentage of immunopositive cells (described earlier), the staining intensity is also evaluated. The scoring of staining intensity is done from 0 to 3 (0: negative; +, weak positive; ++, moderate positive and +++, strong positive). The final immunoscores is calculated either by adding or multiplying the each score.

Conclusions

Immunohistochemistry is a valuable aid in the investigation of numerous infectious diseases. Along with adequate history, clinical and pathological data, it enables the pathologist to make a definitive diagnosis in many cases. It facilitate the specific identification of a disease agent and localisation of the pathogen or specific disease marker within resultant tissue lesions, thereby enabling the distinction between the clinically relevant versus the clinically insignificant presence of some microbes within tissue samples.

References

1. Dagleish MP, Benavides J, Chianini F. (2010). Immunohistochemical diagnosis of infectious diseases of sheep. *Small Ruminant Research* **92**: 19–35.
2. Deborah M, Haines, Edward G, Clark. (1991). Enzyme immunohistochemical staining of formalin-fixed tissues for diagnosis in veterinary pathology. *Can Vet J* **32:** 295-302
3. EduardoJ. Eyzaguirre, Walker David H, ZakiSherif R. (2013). Immunohistology of Infectious Diseases. *Diagnostic Immunohistochemistry***2006**: 43–64.
4. Mehta S, Shelling A, Muthukaruppan A, Lasham A, Blenkiron C, Laking C, Print C. (2010). Predictive and prognostic molecular markers for cancer medicine. *Ther. Adv. Med. Oncol.* **2(2):** 125-148.
5. Meuten DJ (2017). *Tumors in Domestic Animals.* 5thedn. Ames, Iowa, John Wiley and Sons Inc.
6. Ramos-Vara JA. (2005). Technical aspects of immunohistochemistry: a review.*Vet. Pathol.* **42:** 405-426.
7. Reddy GBM and Gajendragad MR (2011). Diagnostic immunohistochemistry: applications and troubleshooting. *In*: *Veterinary Diagnostics Current Trends.*

Gupta RP, Garg SR, Nehra V, Lather D., Editors. Satish Serial Publishing House, Delhi.

8. Renshaw S. (2016). *Immunohistochemistry and Immunocytochemistry: Essential Methods*. 2nd edn. John Wiley and amp; Sons, Ltd.

9. Ruiz FS, Alessi AC, Chagas CA, Pinto GA, Vassallo J. (2005). Immunohistochemistry in diagnostic veterinary pathology: a critical review. *J. Bras. Pathol. Med. Lab.* **41(4):** 263-270.

Chapter 7

Current Trends in Vaccines and Vaccination of Poultry

R. Durairajan

Assistant Professor, Veterinary University Training and Research Centre (TANUVAS), Melmaruvathur – 603 319, Tamil Nadu

Human population is growing every year with 1.30 billion people in our country. In the last 25 years, the egg production has increased to 70 billion from few millions and the broiler production has increased to 3.8 million tons from nowhere. The growth rate of commercial layer is 6-8 percent and commercial broiler is 10-12 percent per annum against the growth of agriculture as whole which is around 2.5 percent. This has largely being achieved through breeding, management, nutrition and availability of good quality animal health products especially vaccines. In fact vaccines have become a common every day part in poultry production; It has not noticed huge losses due to vaccine preventable diseases unlike yesteryears. In fact, poultry diseases cause 10 to 20 per cent economic loss to gross value of production in developed poultry industries and this is likely to be higher in developing countries.

The widespread distribution of Newcastle disease (ND) and the epidemics of avian influenza (AI) that have occurred over the last ten years provide examples of the negative impact of such diseases on the poultry producing sector and on society as a whole (1). Different strategies can be implemented to effectively prevent and control the spread of animal diseases at international, national and farm levels and poultry disease control plans often include the use of vaccination. Their use in poultry production is traditionally aimed at avoiding or minimizing the emergence of clinical disease at farm level and thus increasing production. Vaccines and vaccination programmes vary widely, depending on several local factors (*e.g.* type of production, level of biosecurity, local pattern of disease, status of maternal immunity, vaccines available, costs and potential losses).

The purpose of vaccine in poultry industry is:

1. Prevention of infection or occurrence of fatal diseases and reduction in disease severity.
2. Reduce the diseases causing problem in productivity losses. *e.g.* losses in egg production, meat production, feed-conversion *etc.*
3. To protects offspring by transfer of maternal antibody and public health significance.

First Vaccine Development for Poultry

The first license issued for a poultry product was in 1918 to the University of California, Berkeley, for fowl pox vaccine (2). The list of biological products for poultry grew slowly in the 1920s. However, this began to change with the licensing of laryngotracheitis vaccine in 1933; pigeon pox vaccine in 1939; several Newcastle disease vaccines (inactivated in 1946, Roakin strain in 1948, B1 strain in 1950, and La Sota strain in 1952); and the first bronchitis vaccine in 1953 and for mareks disease vaccine developed in the year 1967 (Mareks HVTFc126) by Witter and associates Rispens. The development and improvement of standards and regulatory requirements to address mycoplasma, leukosis, and to ensure the availability of products that meet appropriate standards of purity, safety, potency, and efficacy are also addressed.

Types of Vaccine

A vaccine is a biological product that can be used to safely induce an immune response that confers protection against infection and/or disease on subsequent exposure to a pathogen. There are different approaches available for designing a vaccine against infectious diseases. These approaches are based on the fundamental information available on pathogenesis and development of immunity against the microbe for which vaccine is being developed.Vaccines are generally classified as live or non-live (sometimes loosely referred to as 'inactivated'). In addition to this several other platforms have been developed over the past few decades, including viral vectors, nucleic acid-based RNA and DNA vaccines, and virus-like particles (4).

Example of Commercially Available Vaccine

1. Killed Vaccines

There are multiple large combination of killed vaccines war available (ND, IB, IBD and bacterial vaccines *etc.*).

A. Chemical Inactivation

By Formalin-inactivated virus and β-propiolactone-inactivated virus are the most important examples of inactivated virus vaccines. They are produced by propagation of the virus in the allantoic fluid of chicken embryos, followed by inactivation with β-propiolactone or formalin.

B. Biological Inactivation

If immunological basis for mono or non-specific immune responses of inactivated vaccines is analyzed, the researchers have related it to the structural alteration of antigens during the process of chemical and physical inactivation, or protein extraction (5). Parry *et al.* (1977) showed that live *E. coli* were superior to heat-killed or formalized organisms at eliciting specific mucosal immunity. It is therefore, expected that a method of inactivation or antigen extraction that does not or minimally alter the antigenic moieties of the organism is the key to developing a cross protective immunizing preparation. Phage lysates provide an option with great potentials. Phage lysates comprise a means of effectively killing bacteria while minimally altering their antigenicity (7). *e.g. E.coli*

2. Live Attenuated Vaccines

Live virus vaccines are prepared using live and weak virus that is still able to cause infection. Most of them are prepared in a lyophilized form that can be stored at low temperature up to one year. There are differences in the level of pathogenicity of live vaccines and some of them are recommended for application of one-day-old chicks and other vaccines for old chickens. Multiples of combinations' available

Examples

1. Avirulent strains: ND B1, F, LaSota, D 58, VG-GA, Ulster, V4, I2, MD-SB1
2. Attenuated or modified: Raokin, RDVK, Gumboro vaccines, IB H120, IB H52
3. Related virus- MD HVT
4. Cell associated virus vaccines- MD HVT/SB1
5. Precocious strains- Coccidiosis vaccines

3. Liposome

There is no liposome vaccine has developed for poultry.

4. Recombinant Vaccine

Several genes from different etiologic agents have been cloned, expressed and purified to be tested as vaccines. There are a variety of expression systems for antigenic protein components, such as bacteria, yeast, mammalian cells and insect cells, in which the DNA encoding the antigenic determinant can be inserted and expressed and stimulating both CD4+ and CD8+ T-cell subsets, and, in some cases, the possibility to be administered orally.

Example

1. Recombinant Viral Vectored vaccine: HVT recombinants, ND, IBD, ILT,AI
2. Recombinant Bacterial Vectored vaccine: None
3. Recombinant subunit vaccine: *E.coli*

Advantage of HVT as Vector

It has safety record and have than 30 years of research as HVT use in poutry vaccine research and pathogenic in poultry bird. Easy to administer *in ovo*/S.C/I.M route. Early then persistent lifelong replication with in a flock and break through maternal derived antibody and protection against oncogenic serotype 1 MDV strain. Large coding capacity of heterologus antigen ~150 kb. Has genomic region with no apparent function to insert genes. Stability of encoded antigens and vectors commercially available for expressing ILT, IBD, AI and ND genes.

Disadvantages

HVT vectored vaccines: Dual vaccines, offer a good protection against the target disease depending on which genes are inserted (slow) but there is evidence that they do not protect properly against Marek disease or against the target disease from which some genes were inserted, such as ILT. **Pox vectored vaccines**: Dual vaccines act fast, but are interfered by pre-existing immunity. **NDV vectored vaccines**: Dual vaccines act very fast but are interfered by maternal antibodies. **FAdV vectored vaccines**: Efficacies in layers are Questionable.

5. Nucleic Acid Vaccines (*mRNA*/DNA)

Work in a different way to other vaccines in that they do not supply the protein antigen to the body. Instead they provide the genetic instructions of the antigen to cells in the body and in turn the cells produce the antigen, which stimulates an immune response. Nucleic acid vaccines are quick and easy to develop, and provide significant promise for the development of vaccines in the future.There are currently **no licenced Nucleic acid vaccines in poultry,** but there are many in development in future.

Factors Affecting Vaccine Protection in Poultry

The level of protection afforded by vaccination is affected by many genetic and environmental factors, including age, maternal antibody levels, prior antigen exposure, vaccine schedule and vaccine dose. Although most of these factors cannot be readily modified, age of vaccination and schedule of vaccination are important and key factors in planning immunization programmes. The vaccine dose is established during early clinical development, based on optimal safety and immunogenicity(8). **Example:** The protection induced by day-old administration of inactivated vaccine (regular or concentrated) in the presence or absence of maternal derived antibody (MDA) to H5N2 Avian Influenza Virus induces poor protection against challenge with H5N1 HPAIV and should not be recommended. Vaccination of MDA-positive chickens at a later age (10 days) seems to be a valuable recommendation, although MDAs may still interfere with vaccination to a lesser extent because they are present up to 3 wk post hatch. Therefore, in areas with high infection pressure, when possible, two vaccinations are recommended for optimal protection. Also, it might be advisable to take into account day-old AI MDA titers when one is determining the optimal age of vaccination (9, 10).

Vaccines Availability

Diseases	*Domestic*	*Imported*
CORE VACCINES		
Ranikhet (Newcastle) disease		
Live	F, B1, LaSota, TANUVAS D58, R_2B, K	LaSota. Clone 30, C131, VH
Inactivated	LaSota	LaSota, Clone 30, Kimber
Infectious Bursal Disease		
Live	Intermediate and Intermediate plus types	Intermediate and Intermediate plus types
Inactivated	Intermediate	Intermediate
Avian Infectious Bronchitis		
Live	H 120, M 41, M 48	H 120, M 41, M 48
Inactivated	M41	M 41 and variants
Fowl Pox- Live	Indian origin	US/European origin
Marek's Disease- Live	HVT (serotype 3)+ SB1 (serotype 2)	HVT+ SB1, HVT+ 301B (serotype 2)
Fowl Adenovirus Inactivated	FAdV- 4	
NON CORE VACCINES		
Fowl Cholera		
Inactivated	A 1	A1, 3, 4
Infectious Coryza	Serovars A, B, C	Serovars A, B, C
Mycoplasma gallisepticum		
Live		6/85, TS 11
Inactivated		S 6, R
Mycoplasma synoviae		MS -H

Diseases	*Domestic*	*Imported*
	Avian REO Virus	
Live		S 1133, 17333, 2708, 3005
Inactivated	S 1133+local strains	
	Avian encephalomyelitis	
Live		Calnek 1143
Inactivated	Calnek 1143	
	Chicken Infectious Anaemia	
Live		24-P4, Cux -1
Inactivated	Local strains	

Vaccination Strategy and Application of Vaccines

Age	*ND Challenges*
0 day	ND Killed
5-7 days	ND Priming
12-14 days	IBD Intermediate plus
18-21 days	ND Booster
28-30 days (optional)	ND Booster

There are three vaccination strategies: routine, emergency and preventive vaccination. After establishing the type of vaccine to be used, the route, method and frequency of administration must be defined, as well as the proper way to combine all these components in the vaccination programme. Vaccine delivery systems significantly influence the outcome of vaccination. An improper vaccine application is considered one of the most common reasons for vaccination programme failure. Various methods of administration can be applied as required by different types of poultry operations (at the hatchery or farm). The choice of method will also depend upon other factors such as the type of production, bird, species, size of the flock, length of the production cycle, general health status, maternal immunity, vaccines to be applied, and costs.

Application of Live/Attenuted Vaccines

Mass Vaccination

The goal of a drinking water vaccination is for each chicken to drink a minimum of one dose of vaccine solution. On the surface this may appear to be the simplest method of live poultry vaccine mass administration; however a drinking water application done correctly is time consuming.

Points to be Considered

1. Water Availability

1. Vaccine should be reconstituted in a sufficient volume of water to cater for a two-hour period. Water requirement in lit = Age in days x no of birds/1000

   e.g. for 1000 birds of 18days we need 18 lit

2. Sufficient drinker space is required to allow free access to the vaccine solution 1.5- 2x. Water consumption during the first three weeks of a chick is erratic, thus walking through the house during the vaccination chasing up inactive chicks, especially along the sides of the house. To further stimulate drinking birds should be deprived of water prior to the vaccination. Usually 1-2 hours of water deprivation is sufficient. Drain the water system completely prior to administration to avoid the first chickens quenching their thirst on standard water.

2. Vaccine Viability

Avoid exposure of vaccine vials and vaccine solution to direct sunlight. To prevent neutralisation of the vaccine virus/bacteria drinking water must have a neutral pH and be free of disinfectants (sodium thiosulphate(@16mg/lit). The addition of skimmed milk or skimmed milk powder to the water 20-30 min prior to reconstitution of vaccine is recommended. (2-6 gm Skim milk powder per lit/2 lit skim milk per 100 lit of water). Use 10 per cent Ice to reach temp 20-25°C. Reconstitute vaccine in a smaller quantity of water (5-10 liters) before addition to drinking water to ensure homogeneous mixture. Vaccine to be consumed with in 2 hrs

3. Assessing Water Intake

To evaluate the uniformity of water intake a blue dye is added to the vaccine solution. Check: Beak/Crop, 90 per cent should have blue colour.

Spray Administration

The objective- to deliver a live antigen on the mucous membrane of upper respiratory tract (nostrils, turbinate, sinuses and trachea) and in the eyes. Use 250-500ml of water for 1000 doses. Temperature of water should be around 20° C and pressure should be minimum 2 bars. Droplet size at the sprayer nozzle level: Coarse spray: >100 -150µ for primary vaccination of ND and IB. Fine spray: 50-100µ for booster vaccinations of ND and IB. Aerosol: 30-50µ for Mycoplasma vaccine. Close the curtain and switch off light spray at the height of 30-40 cm from birds. Curtains should be opened 5-10 min after completion of spray

Application of Inactivated Vaccines

Bring the vaccine bottle to room temperature (20-25°C) prior to vaccinate.

Use a sterilized needle of 12.5mm (1/2"), 20 gauge. Change needle for every 1000 birds in order to avoid injury and contamination of bacteria or viruses. Shake well vaccine bottle prior to vaccination and during vaccination. In case using automatic syringe, periodically check for delivery. Physically check vaccinated birds for deposition vaccine at the correct site. Wet feathers are indicating that vaccine was poorly administered. For subcutaneous method vaccine is injected under the skin of back side of middle neck, the direction of needle should be towards back of the bird

Future Development of Poultry Vaccines

To be considered improved vaccine delivery mechanisms and thermostable vaccines. *e.g.,* In-ovo vaccines- MD, IBD and Edible vaccines. Peptide vaccines is in research and experiments only not for commercial. *e.g.,* Coccidiosis

Conclusion

Immunization protects all the poultry production system from diseases that previously claimed the lives of millions of birds each year, mostly chicks. There are important knowledge gaps and challenges to be addressed. An incomplete understanding of immune mechanisms of protection and the lack of solutions to overcome antigenic variability have hampered the design of effective vaccines against major diseases such as Fowl cholera-IBD and AI. Apart from the scientific challenges, socio-political barriers stand in the way of safe and effective vaccination for poultry birds. Novel delivery platform that eliminate the need for cold chain and routes of administrations have great implication.

References

1. Belova, A.V., Smutka, L. and Rosochatecká, E. (2013). World chicken meat market–its development and current status. *Acta Universitatis Agriculturae et Silviculturae Mendelianae Brunensis* 60: 15-30.
2. David A Espeseth (2013). Early history of regulatory requirements for poultry biological in the United States.*Avian Diseases* 57: 167-171
3. Alexander, D.J., Aldous, E.W. and Fuller, C.M. (2012) The long view: A selective review of 40 years of newcastle disease research. *Avian Pathology* 41: 329-335.
4. Andrew J. Pollard and Else M. Bijker (2020). A guide to vaccinology: from basic principles to new developments. Nature Reviews: *Immunology* 21: 83-100
5. Lauvau, G., Sujata., Vijh Philip., Kong Tiffany., Horng., Kristen Kerksiek., Natalya Serbina., Roman A., Tuma and Eric G. Pamer (2001). Priming of memory but not effector CD8 T cells by a killed bacterial vaccine. *Science.,* 294: 5547

6. Parry, (1997) Intestinal immune response to *E. coli* antigens in the germ-free chicken. *Immunol.,* 32: 731-741.
7. Pasterneck (2009). Method for vaccination of poultry by using Bacteriophage lysate bacterin. US 2009/0297561 A1
8. Schnyder, J. L. *et al.* (2020) Fractional dose of intradermal compared to intramuscular and subcutaneous vaccination—a systematic review and meta-analysis. *Travel. Med. Infect. Dis.* 37: 101868.
9. J. De Vriese, M. Steensels, V. Palya, Y. Gardin, K. Moore Dorsey, B. Lambrecht, S. Van Borm, T. van den Berg (2010). Passive Protection Afforded by Maternally-Derived Antibodies in Chickens and the Antibodies' Interference with the Protection Elicited by Avian Influenza–Inactivated Vaccines in Progeny. *Avian Diseases* 54(1): 246-252.
10. De Jong M. and Bouma A. (2001). – Herd immunity after vaccination: how to quantify it and how to use it to halt disease. *Vaccine,* 19: 2722-2728

Chapter 8

Laboratory Biosafety

H.V. Murugkar

Principal Scientist, National Institute of High Security Animal Diseases, Bhopal, Madhya Pradesh

Laboratory biosafety is the term used to describe the containment principles, technologies and practices that are implemented to prevent unintentional exposure to pathogens and toxins, or their accidental release in the environment. Laboratory biosafety practices are based on the principle of containment of biological agents to prevent exposure to laboratory workers and the outside environment. Primary containment protects the laboratory workers and the immediate laboratory environment from exposure to biological agents. Primary containment is achieved through good microbiological technique and the use of safety equipment and personal protective equipment. Secondary containment protects the environment outside the laboratory, and is provided by facility design and operational procedures. The containment facility of a laboratory is a very crucial component of biosafety principles to prevent the release of infectious agents into the environment. The concept of biosafety levels in the laboratory is primarily based on this concept of containment.

Biosafety Laboratory Design

The facility (laboratory, procedure, or animal room) must operate with inward directional air flow; *i.e.*, air must move from the corridor (area of least contamination) into the room where staff members manipulate the animal or animal by-products (area of greatest contamination). Air exhausted from these rooms cannot be recirculated to other rooms; the air must be discharged to the outside. When there is a very high potential for aerosol generation, additional facility related engineering controls such as sealed penetrations through the walls, double door access to the rooms, or installation of HEPA filters in the exhaust air system may be necessary. A facility can also apply engineering controls locally to reduce aerosol spread. Containment devices (*e.g.*, bonnet tops for cages, ventilated racks) will help keep a potential zoonotic agent in the cage. Staff should use biological safety cabinets when opening the cages, conducting sampling or necropsy activities, or manipulating infectious materials. All biological safety cabinets and the high efficiency particulate air (HEPA) filters in ventilated cage racks need to be certified and leak tested annually. Staff should use special devices such as sealable centrifuge cups, blenders, or homogenizers whenever there is a high potential for creating aerosols of infectious microorganisms. Workers must also rely on personal protective devices, such as respirators, to minimize their exposure to infectious aerosols when containment devices offer insufficient protection.

Risk Assessment for Determination of Bio-containment Level of a Laboratory

Risk assessment is a critical step in the selection of an appropriate containment level for the microbiological work to be carried out. A detailed local risk assessment should be conducted to determine whether work requires containment level 1, 2, 3 or 4 facilities and operational practices (Table 8.1). In addition to the risk group classifications, which are based on the risk factors inherent to the organism, the other factors associated with laboratory operation that should be examined include potential for aerosol generation, quantity of material and concentration of the agent including its stability in the environment (inherent biological decay rate). The type of work proposed (*e.g.*, *in vitro*, *in vivo*, aerosol challenge studies) and the use of recombinant organisms also influence the risk assessment.

Table 8.1. Class of Biosafety Levels Recommended for Handling Infectious Agents

Biosafety Level (BSL)	*Agent Characteristics*	*Practices*	*Facility Requirements*
1	Agents of unknown or minimal potential hazard to laboratory personnel and the environment, or work with defined and characterized strains of viable microorganisms not known to consistently cause disease in healthy adult humans	Standard Microbiological Practices. Disposal of biological waste after autoclaving. No specific safety equipment required.	Open bench top, sink, autoclave; civil structures easy to clean and disinfect with impermeable surfaces.

Biosafety Level (BSL)	Agent Characteristics	Practices	Facility Requirements
2	Agents of moderate potential hazard to personnel and the environment. Primary hazards to personnel working with these agents relate to accidental percutaneous or mucous membrane exposures, or ingestion of infectious materials	BSL-1 practice plus: ☆ Limited access to laboratories ☆ Use of class II biosafety cabinets (BSCs) ☆ Biohazard warning signs ☆ Precautions for handling sharps ☆ Minimize exposure with personal protective equipment (PPE) and aerosol containment devices ☆ Biosafety manual	BSL-1 facilities plus: ☆ Use of air curtains in the laboratory entry or negative ventilation relative to hallway and uncontrolled areas
3	Indigenous or exotic agents that may cause serious or potentially lethal disease as a result of exposure by the inhalation route.	BSL-2 practice plus: ☆ Controlled access ☆ Shower out system ☆ Decontamination of all items including waste before exit ☆ Change of clothing ☆ Decontamination of lab clothing before laundering	BSL-2 facilities plus: ☆ Completely sealed laboratory facility with double door access ☆ Dedicated air handling and effluent treatment system ☆ Double door autoclaves for solid waste disposal ☆ Gradient negative pressure in containment area
4	Dangerous and exotic agents that pose a high individual risk of life-threatening disease, aerosol transmission, or related agents having an unknown transmission risk	BSL-3 practices plus: ☆ All handling of agents in Class III BSCs or Class II BSCs with full-body air-supplied positive pressure suits. ☆ Highly restricted entry ☆ Extensive emergency protocols	BSL-3 facilities plus: ☆ Separate building or isolated zone ☆ Dedicated supply/ exhaust, vacuum, and decontamination systems

In addition to the inherent characteristics of each organism, the containment system includes the engineering, operational, technical and physical requirements for manipulating a particular pathogen. Physical biocontainment facility needs to be supplemented with adequate biosafety practices. Factors like cataloguing of the infectious agent, standardization of the laboratory procedures and practices and defining the safety equipment to handle these infectious agents are taken into account to determine the biosafety level required to handle an individual pathogen so that the personal safety and the environmental contamination issues are not compromised. The recommended biosafety level for an organism represents the conditions under which the agent can normally be handled safely; however, specific circumstances may dictate raising or lowering recommended conditions.

Laboratory Practices and Techniques

Handwashing is an important means of preventing the spread of infectious contaminant. Workers should wash their hands after removing gloves and before leaving the laboratory, procedure room, or animal room. A handwashing sink should be located near the door of the room. Liquid soap dispensers are preferable to soap bars to minimize cross contamination. After thorough washing, workers should dry their hands with disposable paper towels. Eating, drinking, smoking, handling contact lenses, and applying cosmetics should not be permitted in laboratories, nor should any other activities that might involve hand-to-mouth or hand-to-eye contact, such as mouth pipetting. Staff members should perform all manipulations of potentially infectious materials so as to minimize aerosol production. The person who generates infectious waste or contaminates equipment, work surfaces, or other areas is responsible for decontamination before the next person begins work. Chemical disinfection or, preferably, steam autoclaving, is recommended for decontaminating reusable materials before washing. When finished working, staff members must dispose of contaminated waste materials, and package them according to local infectious waste regulations prior to disposal.

Safety Equipment

Many procedures in the laboratory such as centrifuging, mixing, and pipetting that involve high energy tend to produce respirable aerosols that stay airborne for extended periods and are small enough to be inhaled. Low energy procedures including opening containers and streaking plates produce droplets that settle quickly on surfaces, skin, and mucous membranes. All these procedures increase the exposure risk to the workers and may result in serious laboratory association infections. Safety equipment includes biological safety equipment, safety centrifuge cups, and other engineered controls designed to minimize exposure to biological agents. Biological safety cabinets (BSCs) are the most important safety equipment for protection of personnel and the laboratory environment, and most BSCs also provide product protection. Safety equipment is most effective at minimizing exposure when workers are trained in its proper use and the equipment is regularly inspected and maintained.

Personal Protective Equipment (PPE)

Laboratory workers can protect against splashes and splatters by adhering to careful work practices and rigorous use of personal protective devices. Face shields provide protection for eye and mucus membranes. Biological safety cabinets provide near sterile work environments that offer protection to the worker, the materials they are manipulating, and the work area itself. Lab coats or work uniforms will help prevent contamination of street clothes and should be changed whenever visibly soiled. Staff members should autoclave lab coats before disposal or laundering; soiled labcoats should go to on site or professional cleaners only. Latex or vinyl gloves provide barrier protection for hands. Staff must change gloves that are torn or visibly contaminated, and should autoclave them before disposal. Gloves and other protective devices cannot prevent needle sticks or other unintentional injuries caused by sharp instruments, broken glass, *etc.* Self sheathing needles are available, as are other engineered safety devices. Needles must not be bent, cut, or recapped: they must be discarded directly into puncture resistant and leak-proof containers

Decontamination and Disposal of Bio-hazardous Wastes

Complete decontamination and safe disposal of bio-hazardous waste is a crucial biosafety requirement of a bio-containment laboratory. The level of decontamination and waste disposal facility required to handle the bio-waste depends on the biosafety level of the containment laboratory, type of waste being generated, risk group to which the infectious agent belongs, research agenda of the laboratory and the local environmental rules and regulations (CDC, 2002).

Training

All at-risk persons working in a facility should receive appropriate training on that facility's particular biohazards, precautions, and biohazard evaluation procedures. Personnel should receive annual updates and additional training when procedures or policies change. Laboratory workers and animal care personnel should know how to recognize hazard warning signs, to protect themselves and their coworkers against each hazard, and to react properly in the event of emergencies, such as an unintentional biohazard material release. Training should be appropriate for the employee's education, experience, and language skills, and should be performance based to ensure that employees master the skills before encountering a hazard. The facility should incorporate the syllabus of the training programs into its safety manual. Supervisors should assess each employee's biosafety knowledge during the formal training period and later through subsequent regular observation of routine activities. All persons who work in a laboratory bear responsibility to minimize risk of infection through consistent good safe microbiological practices and procedures.

Chapter 9

An Overview of Recent Advances in Animal Disease Diagnosis

Chintu Ravishankar

Associate Professor, Department of Veterinary Microbiology,
College of Veterinary and Animal Sciences, Pookode,
Kerala Veterinary and Animal Sciences University, Wayanad – 673 576, Kerala
e-mail: chintu@kvasu.ac.in

The current scenario of animal disease diagnosis is a far cry from what existed during the major portion of the previous century when classical or traditional microbiological techniques such as isolation and identification of the infection agents, serology *etc.* were in vogue. Though many of the classical techniques are powerful and were used extensively for animal disease diagnosis, they usually suffered from long turnaround times (TAT), from hours to days. Another technical difficulty encountered in these methods was that not all pathogens are easily cultivable. Other limitations of the traditional diagnostic tests include requirements for additional testing and wait times for characterizing detected pathogens (discernment of species, virulence factors, and antimicrobial resistance *etc.*). The serological methods are limited by the cross-reactions between the types of agents and by the fact that some infectious agents have no clearly identified epitopes that are sufficiently specific. In addition, the specific IgM antibodies are detected only in acute phase of infection. Also many of the tests suffered from problems like unavailability of standardized reagents and hence many reagents have to prepared in the lab itself.

However, towards the end of the 20th century the global availability of equipment and chemicals for research increased and so did the capacity to procure it. This led to the possibility of adoption of the latest techniques in labs world over; provided they have the resources for it. This paper examines the major advances made in the field of animal disease diagnosis in the context of microbiology. This is by no means intended to be an exhaustive review of the developments in this field, but rather focuses on advanced techniques that are currently used, or have the capacity for wide spread adoption in the near future. Having said this, it does not imply that the classical techniques of disease diagnosis have been outdated or that they are not in use. In the appropriate diagnostic setting, even basic diagnostic tests such as staining of a blood smear, rapid plate agglutination test *etc.* are still used with great results. Also, though commercial kits and automation are available, the classical tests for isolation and characterization of the microbes are still used.

Toward the last quarter of the 20th century, two techniques, *viz.* enzyme linked immunosorbent assay (ELISA) and polymerase chain reaction (PCR) have revolutionised the field of infectious disease diagnosis. Though not meant to be employed as a disease diagnostic tool, the simplicity of the former test and its amenability to modification, lead to its adoption in the field of animal disease diagnosis. The test could be used for detection of the infectious agent and, if modified suitably, to detect the antibodies against the agent. Monoclonal antibodies improved the specificity of the test, though the cost increased. However, with the practice of mass vaccination for animal diseases, the detection of antibodies as a diagnostic tool for infectious diseases became less sought after. The vaccination antibodies could give false positive results in ELISA, thus confounding the results. Though the DIVA strategy solves this issue to a certain extent, the lack of DIVA kits for many diseases remains a problem.

In the last 20 years, analysis methods based on the nucleic acid amplification (nucleic acid amplification test/NAAT) have been widely used in all life science fields as a new way for the diagnosis of pathogens like virus, bacteria, and parasites. If a search of assay methods in the scientific literature in the National Center for Biotechnology Information (NCBI) PubMed index is carried out, the predominant hits will be on real-time polymerase chain reaction, isothermal amplification and immunoassay. The global veterinary diagnostics market is estimated to expand at a compound annual growth rate of 8.6 per cent from 2016 to 2021, to reach US$ 6.71 billion by 2021. Its market drivers include innovation in new veterinary diagnostics.

As mentioned previously, NAATs have become indispensable in infectious disease diagnosis. The NAAT procedure works by first amplifying – or making many copies of – the pathogens genetic material, if any is present in a specimen. Amplifying those nucleic acids enables NAATs to detect very small amounts of genetic material in a specimen, making these tests highly sensitive for diagnostic purposes.

PCR technology, which is a NAAT, has been widely used to detect pathogenic microorganisms that cause various infectious diseases. This methodology has revolutionized the area of molecular diagnostics because of its high sensitivity of detection and specificity for determination of infectious agents. In addition, there is a reduction in run time and cost over traditional cultivation methods. The other advantages of PCR when compared with diagnostic methods such as serological assays and culture methods are its rapidity, utility, and versatility in clinical laboratory. Though the conventional single target PCR is still the main stay in any molecular diagnostic lab, many modifications of the technique are gaining importance especially due to the need for testing multiple infectious in less TATs.

To reduce cost and to increase the diagnostic capacity of PCR, multiplex PCR (mPCR) is gaining importance. The mPCR refers to the use of different pairs of primers to simultaneously amplify multiple regions of the nucleic acid of the sample with visualization of the amplified products by gel electrophoresis. The use of multiple primer pairs in mPCRs is an innovation that offers significant benefits in cost, time, and exact diagnosis. The main advantage of this technology is to minimize the number of separate reactions, for example, to detect several pathogens at the same time in a single specimen. This technique makes it possible to diagnose several diseases with a single diagnostic test, with sensitivity, specificity and speed. However, it has some limitations such as the nonspecific products generated through primer-primer interactions that may interfere with the amplification of targets, decreasing sensitivity, and selectivity of reactions.

Nowadays clinical diagnostic approaches rely on quantitative PCR (qPCR/ real-time PCR) as a method to detect and quantify infectious agents. Fluorescence chemistry-based methods have revolutionized molecular diagnostics and become the gold standard for viral load quantification and detection of bacterial and viral pathogens. During qPCR, the nucleic acid is amplified until it produces a certain level of signal which is supplied through a DNA intercalating dye or sequence-specific fluorescent probe. The cycle threshold (Cq), defined as the number of amplification cycles required to reach that signal level, is used to calculate the number of target molecules originally present based on a standard curve. In qPCR, the targets are detected in real time from the sealed PCR plate or tube, and there is no post-PCR processing required; therefore, the risk of false-positive results due to amplicon carryover is substantially decreased compared to conventional PCR.

Though, PCR and qPCR are highly sensitive and yield results in a short time, the instrumentation required for carrying out these tests is sophisticated and costly. To take these tests to the point of care (POC), precise thermal regulation and sophisticated instrumentation will be necessary, which can be expensive and difficult to decontaminate between premises. However, attempts have been made to bring these technologies to the field setting with some of the POC technologies being fully automated: integrating nucleic acid extraction, thermal cycling and reporting results to a single instrument for minimal on-site expertise.

Also, alternative PCR chemistries, such as insulated isothermal PCR (iiPCR) and linear-after the-exponential PCR (LATE–PCR) have been explored as potential veterinary diagnostic solutions. In iiPCR, a horizontal temperature gradient is created across a PCR vessel by heating it from the bottom at a fixed temperature, with the top cooled by surrounding air. By removing the requirement for conventional thermal cycling, iiPCR reactions can be performed using relatively simple portable platforms. Lyophilised RT–iiPCR assays have been developed and analysed in laboratory settings for pan-specific detection of FMDV (shown to identify 63 FMDV strains across all seven serotypes), BTV (detecting all 26 known BTV serotypes), and CSFV, all with comparable sensitivity to their respective, laboratory-based q(RT)–PCR equivalents. Furthermore, direct detection of CSFV from serum dilutions and FMDV from vesicular fluid samples has been shown to yield positive results, somewhat negating the requirement for sample preparation and increasing compatibility with field use. Although rapid, with detectable levels of amplicon generated within 30 min, and ease of use for less-skilled operators, assay flexibility is reduced (quantitative interpretation is not possible) in these platforms.

LATE-PCR generates single-stranded products with predictable kinetics for many cycles beyond the exponential phase. LATE-PCR also introduces new probe design criteria that uncouple hybridization probe detection from primer annealing and extension, increase probe reliability, improve allele discrimination, and increase signal strength by 80-250 per cent relative to conventional PCR. These improvements in PCR are particularly useful for real-time quantitative analysis of target numbers in small samples. LATE-PCR is adaptable to high throughput applications in fields such as clinical diagnostics, biodefense, forensics, and DNA sequencing.

Challenges in exploiting PCR for POC diagnostics have resulted in the development of nucleic acid amplification tests based on isothermal principles that are collectively referred to as iNAAT's. Examples of such technologies include: strand-displacement amplification (SDA), helicase-dependent amplification (HDA), recombinase polymerase amplification (RPA) and loop-mediated amplification (LAMP). One advantage of iNAAT's over PCR is that they amplify their targets at a constant temperature, thereby obviating the time-consuming thermocycling steps required for any two- or three-step PCR. As no thermocycling steps are performed, iNAAT reactions are extremely fast (5–20min), resulting in shorter "sample-to-answer" times than conventional PCR. With no thermocycling requirements, amplification reactions can be conducted using simple instruments such as a conventional water bath or heat block – drastically reducing equipment costs. This particular feature has made isothermal amplification systems an attractive platform for POC use and in resource-limited situations.

Next-generation sequencing technologies have taken a quantum jump in the past decade with cost of sequencing decreasing day by day. These technologies are now being applied in the diagnosis and management of bacterial and viral infections and has revolutionized the field of clinical diagnostics. However, a number of drawbacks such as high cost has limited their use in many laboratories, particularly in resource-limited settings. The techniques of viral metagenomics open novel possibilities for the direct comparative analysis of the genetic compositions of various clinical samples and for the detection of "new, emerging viruses." An excellent and very important example for the practical applicability of viral metagenomics is the recent detection of Schmallenberg virus, a novel orthobunyavirus in Europe, with large epizootiological importance.

Metagenomics has the capacity to detect viruses either as single agents or as components in complex infections, such as multifactorial infectious diseases. Considering this wide range of detection, it is important to remember that the discovery of a novel viral genomic sequence by viral metagenomics does not necessarily prove that the detected viruses are the causative agents of the studied diseases or disease complexes. We shall certainly detect not only a wide range of new or so far undetected important pathogens, but also a high number of nonpathogenic, ubiquitous or omni-present infectious agents. Thus, it is absolutely crucial that the detection of various viruses is followed by a range of studies, which investigate the pathogenicity levels and the causative roles of the detected infectious agents. This means that it is not enough to report on metagenomic findings, but viral metagenomics have to be coupled to a range of other complementary investigations, such as attempting the recovery of the agent by virus isolation or by transfection and by performing pathogenicity and epizootiology studies, considering the Koch's postulates. Simultaneously, the outcome of metagenomics can be further strengthened by a range of associated approaches, such as improved bioinformatic analysis and deeper sequencing. Metagenomics approach has been successfully employed to detect pathogens in different animal species; four examples are given below:

1. Pigs with postweaning multisystemic wasting syndrome (PMWS): Using an initial metagenomic approach, two of the collected clinical specimens were investigated for possible viral co-infections. For verification and prevalence estimates of identified viruses, all samples were analysed. Porcine circovirus 2, Torque teno virus, porcine boca-like virus were identified.
2. Shaking mink syndrome (SMS): This neurological disorder affected farmed mink kids in Denmark, as observed in 2000. Brain homogenates from minks affected by SMS were used to reproduce the disease in three healthy individuals. The experimentally infected animals developed the disease; however, the applied conventional methods were unable to detect any infectious agent. These brain samples were then subjected to metagenomic analysis and Astrovirus was identified
3. Nursery and weaned pigs with diarrhoea: Small intestines from a total of 21 piglets (1–2 weeks old) were collected at nine different locations

in Hungary. The samples were examined for viral etiological agents at the OIE CC in Uppsala, by using a two-step approach. First, all samples were pooled and screened for viral sequences using 454-sequencing. The result was then used to identify individuals carrying novel viruses by PCR and each of these samples was deep-sequenced using Ion Torrent technology. Porcine astrovirus, Kobu virus, calcivirus and rotavirus A were identified.

4. Broilers with severe interstitial nephritis: Severe cases of interstitial nephritis in broilers in Hungary made the responsible veterinarian suspect a new variant of infectious bronchitis virus (IBV) in the background. Kidney samples were collected from two affected 10-day-old broilers on a farm and tested by metagenomic analysis at the OIE CC in Uppsala identified Newcastle disease virus.

As mentioned earlier, bacterial culture and identification have been the cornerstone of pathogen identification in the diagnostic microbiology laboratory. With the advent of Sanger sequencing and later, next-generation sequencing, 16S rRNA next-generation sequencing (16SNGS) has been proposed to be a plausible platform for this purpose. Since the 1970s, Carl Woese and others have started to use 16S rRNA gene sequencing for classification of bacteria, exploiting the fact that each bacterial species has its specific 16S rRNA gene sequence. In the 1990s, this technology was put to more routine clinical use for identification of difficult-to-identify bacteria. Not only has16S rRNA gene sequencing been used as the standard of classifying and identifying bacteria, but it has also led to the discovery of an unprecedented number of novel bacterial genera and species. During the process, it was recognised that for some groups of bacteria, 16S rRNA gene sequencing alone may not be discriminative enough for identifying the bacteria. Therefore, other genes or multiple genes were sequenced for this purpose.

Among the various studied gene targets for bacterial identification, 16S rRNA gene has been the most widely used. Therefore, sequencing of 16S rRNA gene is much more reliable and is now considered the gold standard for identification of most bacterial species. Since the 1990s, a large amount of bacterial 16 rRNA gene sequence data have been generated and deposited in various databases, which have enabled matching of query sequences from any bacterial isolates. In the new genomics era, the availability of bacterial genome sequences has also confirmed the representativeness of 16S rRNA gene in comparison to complete genome for bacterial phylogeny studies. 16S rRNA gene sequencing has played a pivotal role for accurate and rapid identification of bacteria for both research and diagnostic purposes. It is most commonly applied for bacterial isolates of uncertain identities from environmental, veterinary to clinical specimens. In clinical microbiology laboratories, 16S rRNA gene sequences can be used to identify most bacterial species including the common Gram-positive and Gram-negative bacteria. However, its usefulness is best highlighted by the ability to identify bacteria with unusual phenotypic profiles and rare or slow-growing bacteria. As a result, diagnosis of infections caused by unusual, rare or slow-growing bacteria has been made possible, allowing better understanding of their clinical disease associations and epidemiology. In some cases, such accurate identification has also enabled decisions on the appropriate antibiotic regimen and infection control procedures. Another advantage is the ability to detect bacteria from direct clinical specimens without pure isolates, as well as nonviable or non-culturable bacteria. Therefore, 16S rRNA gene sequencing has also been found helpful for rapid diagnosis of bacteraemia, culture-negative infections and infections due to non-culturable bacteria. Nevertheless, transition from the traditional methods of identification of bacteria to 16SNGS requires procurement of costly equipment, timely and sustainable maintenance of these platforms, specific facility infrastructure and technical expertise.

Sequence based approach for disease diagnosis has the following advantages:

1. Rapid identification of co-infections, when more than one pathogen is present
2. Identification of novel pathogens, *i.e.*, ones that share little or no genomic structure with known pathogens
3. Outbreak tracebacks pinpointing the geographic locations of origin
4. Monitoring evolution of known pathogens

Another advancement in the field of diagnostics is the introduction of portable antigen/antibody detection systems. Antigen detection has now been incorporated into portable immuno-chromatographic strip tests, also known as antigen-lateral flow devices (Ag–LFDs), which work by binding both viral antigen and antibody-coated detector particles to bands of capturing monoclonal antibody on a membrane. Ag–LFDs are simple to use, highly portable, rapid, disposable and less expensive, making them an ideal assay format for veterinary POC diagnosis. For high-impact livestock viral diseases, there are a number of examples of commercially available Ag–LFDs. They include those developed for the detection of FMD virus and avian influenza virus. Furthermore, FMDV-specific Ag–LFDs are moving towards routine use, appearing in the OIE Terrestrial Manual of 2017 as a recognised assay technology. Prototype assays have also been developed for a number of other diseases, including swine vesicular disease virus, vesicular stomatitis virus (VSV), peste des petits ruminants virus, and serotype-specific detection of FMDV. LFDs can also be developed to detect specific antibody responses to identify animals that have been exposed to pathogens.

Validation studies show that these LFDs can offer a diagnostic sensitivity equivalent to that of a laboratory-based antigen enzyme-linked immunosorbent assay (Ag–ELISA). Furthermore, combining these field-based assays with smartphones has enabled automated readouts and reporting of results for increased functionality. However, in common with other immunoassay formats, Ag–LFDs have lower analytical sensitivity when compared to molecular tests, limiting their application to the acute clinical phase of disease and samples that contain large amounts of intact target organisms. Although sampling multiple affected animals can increase confidence in diagnosis at the herd level, confirming a negative disease status in animals can be difficult.

In addition to the value of LFDs for local testing, recent studies have shown that nucleic acid can be subsequently recovered from them, permitting further downstream molecular characterisation of the agent using polymerase chain reaction and sequencing approaches. These findings are particularly relevant for disease surveillance in remote settings, where centralised diagnostic capacity is often limited.

To conclude, the field of veterinary diagnostics is fast developing with many cutting technologies being employed to ascertain the etiological agent in disease conditions. The modern diagnostic techniques are the outcome of the ever-competitive field of diagnostics, that is driven by market demand. However, the adoption of these technologies should be based on careful assessment of the financial resources available and the place or site of deployment of these tests.

References

1. Anis E, Hawkins IK, Ilha MRS, Woldemeskel MW, Saliki JT, Wilkes RP. Evaluation of Targeted Next-Generation Sequencing for Detection of Bovine Pathogens in Clinical Samples. J Clin Microbiol. 2018 Jun 25; 56(7): e00399-18. doi: 10.1128/JCM.00399-18. PMID: 29695524; PMCID: PMC6018347.

2. Belák S, Karlsson OE, Blomström AL, Berg M, Granberg F. New viruses in veterinary medicine, detected by metagenomic approaches. Vet Microbiol. 2013 Jul 26; 165(1-2): 95-101. doi: 10.1016/j.vetmic.2013.01.022. Epub 2013 Feb 1. PMID: 23428379.

3. Blomström AL. Viral metagenomics as an emerging and powerful tool in veterinary medicine. Vet Q. 2011 Sep; 31(3): 107-14. doi: 10.1080/01652176.2011.604971. Epub 2011 Aug 2. PMID: 22029881.

4. Howson ELA, Soldan A, Webster K, Beer M, Zientara S, Belák S, Sanchez-Vizcaino JM, Van Borm S, King DP, Fowler VL. Technological advances in veterinary diagnostics: opportunities to deploy rapid decentralised tests to detect pathogens affecting livestock. Rev Sci Tech. 2017 Aug; 36(2): 479-498. doi: 10.20506/rst.36.2.2668. PMID: 30152469.

5. Kubacki J, Fraefel C, Bachofen C. Implementation of next-generation sequencing for virus identification in veterinary diagnostic laboratories. J Vet Diagn Invest. 2021 Mar; 33(2): 235-247. doi: 10.1177/1040638720982630. Epub 2020 Dec 24. PMID: 33357110; PMCID: PMC7953084.

6. Kumar, D., Rao, P.P. and Hegde, N.R. 2019. Next-generation sequencing as diagnostic tool in veterinary research. J. Anim. Res., 9(6): 797-806
7. Kwok KTT, Nieuwenhuijse DF, Phan MVT, Koopmans MPG. Virus Metagenomics in Farm Animals: A Systematic Review. Viruses. 2020 Jan 16; 12(1): 107. doi: 10.3390/v12010107. PMID: 31963174; PMCID: PMC7019290.
8. Lau, S.K.P., Teng, J.L.L., Ho, C-C, Woo P.C.Y. Gene Amplification and Sequencing for Bacterial Identification in *Methods in Microbiology*, Sails, A., Tang Y-W (Eds), Academic Press, Volume 42, 2015, Pages 433-464, https://doi.org/10.1016/bs.mim.2015.04.003
9. Obande GA, Banga Singh KK. Current and Future Perspectives on Isothermal Nucleic Acid Amplification Technologies for Diagnosing Infections. Infect Drug Resist. 2020 Feb 12; 13: 455-483. doi: 10.2147/IDR.S217571. PMID: 32104017; PMCID: PMC7024801.
10. OIE Bulletin. High-throughput sequencing in veterinary infection biology and diagnostics. OIE Bulletin 2013-2. Progress in technology. https://www.oie.int/fileadmin/Home/eng/Publications_and_Documentation/docs/pdf/bulletin/Bull_2013-2-ENG.pdf (Accessed 26 July 2021)
11. Perlejewski K, Bukowska-Ośko I, Rydzanicz M, Pawełczyk A, Caraballo Cortès K, Osuch S, Paciorek M, Dzieciątkowski T, Radkowski M, Laskus T. Next-generation sequencing in the diagnosis of viral encephalitis: sensitivity and clinical limitations. Sci Rep. 2020 Sep 30;10(1):16173. doi: 10.1038/s41598-020-73156-3. PMID: 32999423; PMCID: PMC7528011.
12. Resende TP, Marshall Lund L, Rossow S, Vannucci FA. Next-Generation Sequencing Coupled With *in situ* Hybridization: A Novel Diagnostic Platform to Investigate Swine Emerging Pathogens and New Variants of Endemic Viruses. Front Vet Sci. 2019 Nov 15; 6: 403. doi: 10.3389/fvets.2019.00403. PMID: 31803766; PMCID: PMC6873589.
13. Sanchez JA, Pierce KE, Rice JE, Wangh LJ. Linear-after-the-exponential (LATE)-PCR: an advanced method of asymmetric PCR and its uses in quantitative real-time analysis. Proc Natl Acad Sci U S A. 2004 Feb 17; 101(7): 1933-8. doi: 10.1073/pnas.0305476101. Epub 2004 Feb 9. Erratum in: Proc Natl Acad Sci U S A. 2009 Feb 10; 106(6): 2083. PMID: 14769930; PMCID: PMC357030.
14. Souf S. Recent advances in diagnostic testing for viral infections, Bioscience Horizons: The International Journal of Student Research, Volume 9, 2016, hzw010, https://doi.org/10.1093/biohorizons/hzw010
15. Suchodolski JS. Diagnosis and interpretation of intestinal dysbiosis in dogs and cats. Vet J. 2016 Sep; 215: 30-7. doi: 10.1016/j.tvjl.2016.04.011. Epub 2016 Apr 25. PMID: 27160005.

16. Van Borm S, Belák S, Freimanis G, Fusaro A, Granberg F, Höper D, King DP, Monne I, Orton R, Rosseel T. Next-generation sequencing in veterinary medicine: how can the massive amount of information arising from high-throughput technologies improve diagnosis, control, and management of infectious diseases? Methods Mol Biol. 2015; 1247: 415-36. doi: 10.1007/978-1-4939-2004-4_30. PMID: 25399113; PMCID: PMC7123048.

Chapter 10

Nobel Approaches to Detect Biofilm Forming and ESBL Producing Bacteria of *Enterobacteriaceae* Family from Livestock and Poultry

Indranil Samanta

Department of Veterinary Microbiology,
West Bengal University of Animal and Fishery Sciences, Kolkata
e-mail: isamanta76@gmail.com

Introduction

Human and animal population are at risk of cross-transmission of zoonotic bacteria via direct contact due to close proximity with food animals, companion animals, live wildlife market, environmental contamination, and intake of contaminated animal origin food items. The situation becomes more complicated due to cross-transmission of antimicrobial resistance (AMR) determinants along with the infection. In human, over prescription of antimicrobials is the major source of selection for antimicrobial resistant bacteria but use of antimicrobials in food animals, and moreover, environmental resistance gene pool ('resistome') also play important roles in this complex multi-factorial state of affairs.

AMR leads to enormous financial losses associated with high morbidity, mortality, man-day loss, hospital length of stay (LOS), direct health-care costs and social costs of infection (Wozniak *et al.*, 2019). About 700,000 deaths per year were attributed to AMR alone that is more than the toll caused by malaria, acquired immunodeficiency syndrome (AIDS) and tuberculosis (O'neill, 2014). World health organization (WHO) identified eight pathogens relevant to AMR, among them five are pathogenic or commensal bacteria (*Klebsiella pneumoniae,*

Escherichia coli, Staphylococcus aureus, Neisseria gonorrhoeae, Mycobacterium tuberculosis) (WHO, 2018). The third-generation cephalosporin-resistant and carbapenem-resistant *Enterobacteriaceae* (CRE, *e.g. Escherichia coli* and *Klebsiella pneumoniae*) alone were reported to cause 6.4 million bloodstream infections and 50.1 million serious infections worldwide in a year (Temkin *et al.*, 2018). A recent estimate suggested 33,000 annual deaths due to AMR in European Union and European Economic Area (Cassini *et al.*, 2019). Additional treatment cost and loss due to methicillin-resistant *Staphylococcus aureus* (MRSA), third-generation cephalosporin-resistant and ESBL-producing *Enterobacteriaceae* ranged between 1732-9726 USD and 2.54-6.8 days per case, respectively (Wozniak *et al.*, 2019). For the United States alone, average national health care expenditure was estimated at around 2.2 billion US Dollar due to AMR (Thorpe *et al.*, 2018).

Addressing AMR in developing countries was considered crucial by the United Nations to attend sustainable development goals (SDGs) associated with poverty and hunger alleviation and improvement of health and economic growth. In low-and-middle-income countries (LMICs) such as India, the current rate of AMR related infections is high and is projected to grow more rapidly than the developed countries. A substantial portion (40-60 per cent) of human bacterial infections in Brazil, the Russian Federation and India is associated with resistant bugs in comparison to developed countries (17 per cent) (OECD, 2018). In LMICs the direct and prominent effects of AMR includes increased mortality in addition to higher morbidity and economic losses (Okeke *et al.*, 2005). The recent projection about the financial vulnerability of LMICs revealed that additional 19 million people are going to fall into great poverty by 2030 due to AMR producing direct impacts on labour productivity (neat GDP produced by one hour labour) and increased health care costs (Ahmed *et al.*, 2017).

Exposure of Livestock to the Antimicrobials

Rearing of food animals including poultry is dependent on usage of medically important antibiotics (MIA) for therapy, prophylaxis and growth promotion. Recently world health organization (WHO) recommended complete cessation of MIA use in healthy animals for prophylaxis or growth promotion (WHO, 2019). The European Centre for Disease Prevention and Control, European Food Safety Authority and European Medicines Agency joint report revealed more average annual antibiotic consumption in animals than human (ECDC/EFSA/EMA, 2017). It is estimated that two-third of the antibiotics produced is consumed by the livestock sector globally (CDDEP, 2015), which is expected to increase by 67 per cent in 2030 specially in BRICS group of countries including India to meet the increasing animal protein demand (Van Boeckel *et al.*, 2015). The data for antimicrobial usage in livestock in LMICs including India is scarce but it is evident that use of antibiotics varies with the farming practices. The backyard poultry farmers mostly use ethno-veterinary medicine based on indigenous technical knowledge and in more severe cases with the lower generation and cheaper antibiotics available in the local retail shop (Samanta *et al.,* 2018). The antimicrobial usage for growth promotion of livestock and poultry was not detected among the Indian backyard farmers.

Other than direct way the livestock are also exposed to the antimicrobials by means of indirect ways such as exposure to the contaminated environment. Antibiotics treated human or animal waste, effluents from antimicrobial production units are found responsible for contamination of the environment with the resistant bugs or the 'resistome'. In India, effluent from water-treatment plant was found to contain high concentration of fluroquinolone (ciprofloxacin) causing contamination of surface, ground, and drinking water (Larsson *et al.,* 2007). Our studies found more exposure to the environment presumed to be contaminated might be the probable reason for higher occurrence of beta-lactamase producing bacteria in backyard pigs, kuroilers and ducks than the farmed counterpart, although not directly exposed to the antimicrobials (Samanta *et al.,* 2015; Mahanti *et al.,* 2018; Banerjee *et al.,* 2019).

Transmission

The transmission of antimicrobial resistance from animals to humans takes place through contaminated animal origin food, direct contact between animals and humans, and release of antibiotic residues from food-animals and birds into the environment mainly through the manure.

Extended-Spectrum-Beta-Lactamase Producing *Enterobacteriaceae*

The Gram-negative bacteria (GNB) which are cephalosporin-resistant mainly produce the resistance through generation of β-lactamase or cephalosporinase such as AmpC type β-lactamase (ACBL) and extended-spectrum β-lactamase (ESBL) which have become a cause of concern for their global spread, high infectivity and associated mortality. Food and companion animals being the reservoir of ESBL or ACBL producers can put persons in their contact or consumers at significant risk. Due to abysmal public health infrastructure and poor hygiene, LMICs are overburdened with neonatal sepsis and health-care associated infections with a heavy load of such drug-resistant pathogens in hospitals and community (Kang *et al.,* 2012). About 22 per cent of healthy humans from Southeast Asia were found to harbour ESBL-producing bacteria in their gut-which is much higher than the global average (14 per cent) (Karanika *et al.,* 2016). Increased contact with food animals is often assumed as an important underlying determinant for such higher ESBL colonization in Asian and African population (Miao *et al.,* 2017). The number of studies from LMICs implicated food-animals as important reservoirs of ESBLs, however, their role in human infection cannot be ascertained. In contrast, chickens were reported to acquire ESBL resistance from water contaminated with human sewage through integrated farming which is a common practice in many parts of the Southeast Asia (Nguyen *et al.*, 2015). In bovine, ESBL-producers with multiple CTX-M variants were reported from mostly fecal sources and mastitic milk with a preponderance of CTXM-1 and CTXM-15 variants. Among other common types of the ESBLs in bovine from various LMICs were SHV-180, OXA-10, SHV-5 (Turkey), SHV-12 (Turkey; Egypt), SHV-180 (India), OXA-30 (Egypt) and TEM (India) (DantasPalmeira and Ferreira, 2020). A number of studies conducted by our group revealed the presence of some ESBL variants in bovine mastitis wherein we could not trace any known ESBL genes by PCR screening – existence of a novel ESBL mechanism in this part cannot be ruled out (Koopvara *et al.*, 2016; Bandyopadhyay *et al.*, 2018). Apart from a major source of animal protein, cattle manure is widely used as biological fertilizer which increases the chance of environmental dissemination of ESBLs from fecal source. Separate studies conducted by our group revealed the presence of ESBLs in broilers or birds kept exclusively for meat (Mahanti *et al.*, 2018) but not in backyard poultry reared by a large section of resource poor farmers in India (Samanta *et al.*, 2014). Importantly, pigs which forms a favorite and affordable dish in many protein hungry countries were reported to carry ESBL determinants (Samanta *et al.*, 2015; 2018).

Detection Techniques for ESBL

For the detection of ESBLs and AmpC β-lactamases, phenotypic and genotypic methods are employed. Phenotypic approaches to detect ESBL and/or AmpC are based on the detection of synergy between β-lactam agents and specific substances that inhibit each enzyme type. In the broth microdilution (BMD), synergy is indicated by a substantial decrease of the minimum inhibitory concentration (MIC) of the β-lactam. Following this principle, several methods that employ disks containing β-lactams and inhibitors have been validated and are widely used (CLSI, 2017). Since these assays require overnight incubation, the turnaround time amounts to 18 hours. This delay is also the main disadvantage of other culture-based approaches such as the double-disk synergy test, three-dimensional tests, gradient diffusion tests as well as automated systems. Rapid testing with disk diffusion has been described, however, being recommended only for preliminary susceptibility reports. It requires further standardization and adapted clinical breakpoints for a correct interpretation. Colorimetric methods represent a substantially faster alternative, although some of them display a low positive predictive value in the presence of AmpC hyperproduction (Decousser *et al.*, 2017).

Methods based on matrix-assisted laser desorption ionization time-of-flight mass spectrometry (MALDI-TOF MS) can also be applied to identify β-lactam resistance. One of these approaches relies on the detection of products resulting from the hydrolysis of β-lactam antibiotics by β-lactamases. Nonetheless, this depends on several factors that are specific to the enzyme-substrate interaction (such as enzyme availability), which could interfere with the results, as well as alternate mechanisms that also lead to misdetection of antibiotic hydrolysis. Moreover, a positive result indicates the presence of resistance, but does not provide an exact MIC value for the tested antibiotic (Decousser *et al.*, 2017). Another strategy consists in detecting isotopically labeled amino acids, which requires a specific culture medium that can be supplemented with these substances. The applicability of techniques that analyze the amount of biomass resulting from bacterial incubation with antibiotic agents in order to establish susceptibility patterns has also been described (Sparbier *et al.*, 2016).Considering the need for rapid methods that are easy to standardize, a screening panel for detection of ESBL and AmpC β-lactamases in *Enterobacterales* adapting the principle of the MALDI-TOF MS-based direct-on-target microdroplet growth assay (DOT-MGA) was developed recently (Correa-Martínez *et al.*, 2019).

Detection of Biofilm Producing Bacteria

Biofilms are communities of bacteria attached irreversibly with organs or polystyrene which showed increased resistance against host immune response. The environment and putative factors such as *rpoS*, *sdiA*,*rcsA* and others regulate the transformation of planktonic cells into the biofilm (Ranjit and Young, 2013). Biofilm formation in bacterial isolates is a protective response to the external stresses including antibiotics for better survival (Kaplan, 2011). Biofilm producing bacteria are detected to be 1000 times more resistant to different antibiotics. The comparative study detected the tissue culture plate (TCP) method as a more quantitative and reliable method for the detection of biofilm forming microorganisms than tube method (TM) and Congo Red agar (CRA) methods in the laboratories (Hassan *et al.*, 2011). Our studies detected occurrence of moderate or strong biofilm producing capacity by TCP method in *E. coli/Klebsiella* isolates present in healthy companion/backyard animals and ducks (Banerjee *et al.*, 2019; 2020).

References

1. Ahmed, S.A., Baris, E., Go, D.S., Lofgren, H., Osorio-Rodarte, I., Thierfelder, K., 2017. Assessing the global economic and poverty effects of antimicrobial resistance. Policy Research Working Paper: 8133. Available at: https: //openknowledge.worldbank.org/handle/10986/27636; accessed on 11/06/2021

2. Bandyopadhyay, S., Banerjee, J., Bhattacharyya, D., Samanta, I., Mahanti, A., Dutta, T.K., Ghosh, S., Nanda, P.K., Dandapat, P. and Bandyopadhyay, S., 2018. Genomic identity of fluoroquinolone-resistant bla CTX-M-15-Type ESBL and pMAmpC β-lactamase producing Klebsiella pneumoniae from buffalo milk, India. *Microbial Drug Resistance*, 24(9), pp.1345-1353.

3. Banerjee, A., Bardhan, R., Chowdhury, M., Joardar, S.N., Isore, D.P., Batabyal, K., Dey, S., Sar, T.K., Bandyopadhyay, S., Dutta, T.K. and Samanta, I., 2019. Characterization of beta-lactamase and biofilm producing

Enterobacteriaceae isolated from organized and backyard farm ducks. *Letters in applied microbiology, 69*(2), pp.110-115.

4. Banerjee, A., Batabyal, K., Singh, A.D., Joardar, S.N., Dey, S., Isore, D.P., Sar, T.K., Dutta, T.K., Bandyopadhyay, S. and Samanta, I., 2020. Multidrug resistant, biofilmproducing highrisk clonal lineage of Klebsiella in companion and household animals. *Letters in Applied Microbiology, 71*(6), pp.580-587.
5. Cassini, A., Högberg, L.D., Plachouras, D., Quattrocchi, A., Hoxha, A., Simonsen, G.S., Colomb-Cotinat, M., Kretzschmar, M.E., Devleesschauwer, B., Cecchini, M. and Ouakrim, D.A., 2019. Attributable deaths and disability-adjusted life-years caused by infections with antibiotic-resistant bacteria in the EU and the European Economic Area in 2015: a population-level modelling analysis. *The Lancet infectious diseases, 19*(1), pp.56-66.
6. Centre for Disease Dynamics, Economics and Policy (CDDEP), 2015. State of the World's Antibiotics, Washington, DC: CDDEP.
7. CLSI, 2017. Clinical and Laboratory Standards Institute: Performance standards for antimicrobial susceptibility testing: Twenty-seventh informational supplement, M100, Wayne, PA; 2017.
8. Correa-Martínez, C.L., Idelevich, E.A., Sparbier, K., Kostrzewa, M. and Becker, K., 2019. Rapid detection of extended-spectrum β-lactamases (ESBL) and AmpC β-lactamases in Enterobacterales: development of a screening panel using the MALDI-TOF MS-based direct-on-target microdroplet growth assay. *Frontiers in microbiology, 10*, p.13.
9. DantasPalmeira, J. and Ferreira, H.M.N., 2020. Extended-spectrum beta-lactamase (ESBL)-producing Enterobacteriaceae in cattle production—a threat around the world. Heliyon 6, e03206.
10. Decousser, J.W., Poirel, L. and Nordmann, P., 2017. Recent advances in biochemical and molecular diagnostics for the rapid detection of antibiotic-resistant Enterobacteriaceae: a focus on β-lactam resistance. *Expert review of molecular diagnostics, 17*(4), pp.327-350.
11. Hassan, A., Usman, J., Kaleem, F., Omair, M., Khalid, A. and Iqbal, M., 2011. Evaluation of different detection methods of biofilm formation in the clinical isolates. *Brazilian journal of infectious Diseases, 15*(4), pp.305-311.
12. Kang, C.I., Wi, Y.M., Lee, M.Y., Ko, K.S., Chung, D.R., Peck, K.R., Lee, N.Y. and Song, J.H., 2012. Epidemiology and risk factors of community onset infections caused by extended-spectrum β-lactamase-producing Escherichia coli strains. *Journal of clinical microbiology, 50*(2), pp.312-317.
13. Kaplan, J.B., 2011. Antibiotic-induced biofilm formation. *The International journal of artificial organs, 34*(9), pp.737-751.

14. Karanika, S., Karantanos, T., Arvanitis, M., Grigoras, C. and Mylonakis, E., 2016. Fecal colonization with extended-spectrum beta-lactamase–producing Enterobacteriaceae and risk factors among healthy individuals: a systematic review and metaanalysis. *Reviews of Infectious Diseases, 63*(3), pp.310-318.

15. Koovapra, S., Bandyopadhyay, S., Das, G., Bhattacharyya, D., Banerjee, J., Mahanti, A., Samanta, I., Nanda, P.K., Kumar, A., Mukherjee, R. and Dimri, U., 2016. Molecular signature of extended spectrum β-lactamase producing Klebsiella pneumoniae isolated from bovine milk in eastern and north-eastern India. *Infection, Genetics and Evolution, 44*, pp.395-402.

16. Larsson, D.J., de Pedro, C. and Paxeus, N., 2007. Effluent from drug manufactures contains extremely high levels of pharmaceuticals. *Journal of hazardous materials, 148*(3), pp.751-755.

17. Mahanti, A., Ghosh, P., Samanta, I., Joardar, S.N., Bandyopadhyay, S., Bhattacharyya, D., Banerjee, J., Batabyal, S., Sar, T.K. and Dutta, T.K., 2018. Prevalence of CTX-M-Producing Klebsiella spp. in broiler, kuroiler, and indigenous poultry in West Bengal State, India. *Microbial Drug Resistance, 24*(3), pp.299-306.

18. Miao, Z., Li, S., Wang, L., Song, W. and Zhou, Y., 2017. Antimicrobial resistance and molecular epidemiology of ESBL-producing Escherichia coli isolated from outpatients in town hospitals of Shandong Province, China. *Frontiers in microbiology, 8*, p.63.

19. Nguyen VT, Carrique-Mas JJ, Ngo TH, Ho HM, Ha TT, Campbell JI, Nguyen TN, Hoang NN, Pham VM, Wagenaar JA, Hardon A. Prevalence and risk factors for carriage of antimicrobial-resistant Escherichia coli on household and small-scale chicken farms in the Mekong Delta of Vietnam. *J Antimicrob Chemother* (2015) 70: 2144-2152.

20. O'neill, J., 2014. Antimicrobial resistance: tackling a crisis for the health and wealth of nations. *Rev Antimicrob Resist* 20, 1-16.

21. OECD., 2018. Stemming the Superbug Tide: Just A Few Dollars More, OECD Publishing, Paris. https: / / doi.org / 10.1787 / 9789264307599-en. accessed on 11 / 06 / 2021.

22. Okeke, I.N., Laxminarayan, R., Bhutta, Z.A., Duse, A.G., Jenkins, P., O'Brien, T.F., Pablos-Mendez, A. and Klugman, K.P., 2005. Antimicrobial resistance in developing countries. Part I: recent trends and current status. *The Lancet infectious diseases, 5*(8), pp.481-493.

23. Ranjit, D.K. and Young, K.D., 2013. The Rcs stress response and accessory envelope proteins are required for de novo generation of cell shape in Escherichia coli. *Journal of bacteriology, 195*(11), pp.2452-2462.

24. Samanta, A., Mahanti, A., Chatterjee, S., Joardar, S.N., Bandyopadhyay, S., Sar, T.K., Mandal, G.P., Dutta, T.K. and Samanta, I., 2018. Pig farm

environment as a source of beta-lactamase or AmpC-producing Klebsiella pneumoniae and Escherichia coli. *Annals of Microbiology*, *68*(11), pp.781-791.

25. Samanta, I., Joardar, S.N., Das, P.K., Das, P., Sar, T.K., Dutta, T.K., Bandyopadhyay, S., Batabyal, S. and Isore, D.P., 2014. Virulence repertoire, characterization, and antibiotic resistance pattern analysis of Escherichia coli isolated from backyard layers and their environment in India. *Avian diseases*, *58*(1), pp.39-45.
26. Samanta, I., Joardar, S.N. and Das, P.K., 2018. Biosecurity strategies for backyard poultry: a controlled way for safe food production. In: *Food control and biosecurity* (pp. 481-517). Academic Press.
27. Samanta, I., Joardar, S.N., Mahanti, A., Bandyopadhyay, S., Sar, T.K. and Dutta, T.K., 2015. Approaches to characterize extended spectrum beta-lactamase/beta-lactamase producing Escherichia coli in healthy organized vis-a-vis backyard farmed pigs in India. *Infection, Genetics and Evolution*, *36*, pp.224-230.
28. Sparbier, K., Schubert, S. and Kostrzewa, M., 2016. MBT-ASTRA: a suitable tool for fast antibiotic susceptibility testing?. *Methods*, *104*, pp.48-54.
29. Temkin, E., Fallach, N., Almagor, J., Gladstone, B.P., Tacconelli, E., Carmeli, Y. and DRIVE-AB Consortium, 2018. Estimating the number of infections caused by antibiotic-resistant Escherichia coli and Klebsiella pneumoniae in 2014: a modelling study. *The Lancet Global Health*, *6*(9), pp.e969-e979.
30. Thorpe, K.E., Joski, P. and Johnston, K.J., 2018. Antibiotic-resistant infection treatment costs have doubled since 2002, now exceeding $2 billion annually. *Health Affairs*, *37*(4), pp.662-669.
31. Van Boeckel, T.P., Brower, C., Gilbert, M., Grenfell, B.T., Levin, S.A., Robinson, T.P., Teillant, A. and Laxminarayan, R., 2015. Global trends in antimicrobial use in food animals. *Proceedings of the National Academy of Sciences*, *112*(18), pp.5649-5654.
32. WHO., 2018. Antimicrobial resistance. Available at: https: //www.who.int/news-room/fact-sheets/detail/antimicrobial-resistance (2018); accessed on 11/06/2021
33. WHO., 2019. WHO guidelines on use of medically important antimicrobials in food producing animals. Rep., WHO, Geneva. Available at: https: //www.who.int/foodsafety/areas_work/antimicrobial-resistance/cia_guidelines/en/; accessed on 11/06/2021
34. Wozniak, T.M., Barnsbee, L., Lee, X.J. and Pacella, R.E., 2019. Using the best available data to estimate the cost of antimicrobial resistance: a systematic review. *Antimicrobial Resistance and Infection Control*, *8*(1), pp.1-12.

Chapter 11

Vaccine against Poultry Coccidiosis: A Cup Half Full?

Vivek Agrawal, Amit Kumar Jaiswal, Anantrao Kashiram Jayraw, Mukesh Shakya and Nirmal Jamra

Department of Veterinary Parasitology,
College of Veterinary Science and Animal Husbandry, Mhow
Nanaji Deshmukh Veterinary Science University, Jabalpur, Madhya Pradesh

Introduction

Losses caused by intestinal coccidiosis (*Eimeria*) in poultry industry are well recognized from more than a century (Chapman, 2014) and their imperative informations like life cycle, pathogenesis, global distribution are now well established (Shirley *et al.*, 2005). A heavy infection of coccidiosis in poultry leads to necrotic enteritis, reduced feed conversion ratio and even some species of *Eimeria* are responsible for mortality of birds (Shirley *et al.*, 2005; Williams, 1999). Due to heavy economic losses to poultry industry by *Eimeria*, prophylactic use of anticoccidial drugs and parasitic vaccine are appear in light (Blake and Tomley, 2014).

Anticoccidial drug sensitivity test is only available tool to diagnose the anticoccidial drug resistance (Naciri *et al.*, 2003) nevertheless it has so many inherent limitation like expensiveness, isolates frequently originate from disease outbreaks (and may not always be representative for the field) and need to be propagated first in specified pathogen-free chickens for multiplication (which may result in the selection of non-relevant coccidia). Unlike other pathogens, no major studies has been materilize at molecular level in case of *Eimeria* due to its complex structure and mechanism involved. Otherwise it is imperative to know the exact mechanism behind the development of resistance against anticoccidial drugs as well as vaccine.

Allelic diversity of selected gene encoding antigens used to make recombinant vaccine and play a seminal role if allelic diversity is more. The chances of failure of vaccine is due to immune escape by various protozoans (*Plasmodium falciparum* with the antimalarial vaccine candidates apical membrane antigen 1 (AMA1) and merozoite surface antigen 1 (MSP1) (Arnott *et al.*, 2014; Takala and Plowe, 2009) but on the other hand selection pressure on recombinant antigen is less and thus the chances of development of resistance against recombinant vaccine is also limited.

Eimeria show very high selection pressure due to mainly peculiarity of parasite as well as host factors fecundity, replication of parasite, high number of host,their intense farming and rapid turnover are the determinants for the development of anticoccidial drug resistance (Chapman, 1997). In a live vaccine, whole parasite is used which shows 6000-9000 antigens during their completion of life cycle (Reid *et al.*, 2014). Therefore till date even after the use of 50 years, no resistance has been reported against live vaccine but situation is different for recombinant vaccine because few antigens are used then there are more chances immune evasion or resistance by parasite might be occurred owing to more focused genetic selection pressure.

Salient Feature Pertaining to Coccidiosis

Coccidiosis caused by *Eimeria* protozoa belong to phylum Apicomplexa and is very close relative of *Toxoplasma gondii* and *Plasmodium* spp. *Eimeria* is having capacity to infect all livestock including poultry but most of the species are host specific (Vrba and Pakandl, 2015). The global economic loss to poultry industry by this apicomplexa is approximately £2 billion (USD 2750 million) per annum (Lassen and Ostergaard, 2012). The estimated annual loss of USD 140 million to Indian poultry industry in terms of high morbidity and reduced waight gain also attributed to coccidiosis. There are seven species of *Eimeria viz.*; *Eimeria tenella, Eimeria necatrix, Eimeria brunetti, Eimeria acervulina, Eimeria maxima, Eimeria mitis, Eimeria praecox* are responsible for clinical coccidiosis manifested in form of haemorrhagic or malabsorptive enteritis (Williams *et al.*, 2009).

The limitation with anticoccidial drugs for poultry *Eimeria* is residual effect on poultry products and hence countries like USA and European union has banned it for prophylaxis.

Hence researcher and academia both are actively involved to find out golden bullet or cocktail of antigens against closely related parasite like *Toxoplasma gondii* and *Plasmodium* spp. but they did not get it owing to high standard (Gedik *et al.*, 2016). Same things might be true with *Eimeria* spp. also.

Leakiness of *Eimeria* Oocyst

Leakiness pays a key role during use of ionophores and there are multifaceted advantages of that phenomena *viz.* few *Eimeria* oocyst remain unexposed leads to development of immunity owing to trickle infection and there is low selection pressure on *Eimeria* oocyst which helps to slow development of resistance. Owing to a aforementioned rationale, ionophores are the anticoccidial drugs which share more than 70 per cent of anticoccidial drug market globally (Eckford *et al.*, 2014), although due to health concern issue Codex put maximum residue limit of anticoccidial in poultry meat and accordingly. European union and United states have curtain the use of anticoccidial and encouraging the use of vaccine for control of economically significant coccidiosis. Nonetheless after introduction of ionophores, resistance against them documented might be due to pre exiting genetic diversity of *Eimeria* spp. or high genome plasticity.

Vaccine Development - So far Study

The concept of vaccine against *Eimeria* came in existence when poultry become resistant to further infection of *Eimeria* after ingestion of homologous *Eimeria* spp. This was the basis for modern coccidiology vaccine. The first vaccine developed against *Eimeria* was live vaccine (Coccivac). The inherent limitation of live vaccine was to develop the lesion in intestine and leading to subclinical coccidiosis. Another major constraint is cost of production (Blake and Tomley, 2014). These limitation resulted the development of attenuated vaccine which is having upper hands as far as the production of lesion is concerned which is absent in case of attenuated vaccine (Paracox) and their low cost of production (Chapman and Jeers, 2014) but at the same time by using the attenuated vaccine, we lost the brighter site of live vaccine that is reversal of susceptibility of anticoccidial drug against *Eimeria* (Chapman and Jeers, 2015). In Europe live vaccine is not currently licenced due to their vaccine induced drawback. Production of both type of vaccine at mass level is major hindrance somewhere and immunity achieved during the attenuated vaccine was not up to desired level therefore there is a need for the development of subunit vaccine.

The first subunit vaccine was CoxAbic used commercially at broad scale. It is based on gam56 and gam82 purified gametocyte of *E. maxima*. It is vital antigen involved in synthesis of oocyst wall due to which it become resistant against adverse environmental condition and transmission of occyst occur very quickly. As there is protein cross linked via ditrypsin bond required during cell wall synthesis. CoxAbicprovide the protection upto 8 week in broiler chick which is the age to sale in market. Another key feature of CoxAbic is to provide immunity against multiple species of *Eimeria* (Smith, 1994) unlike species specific direct immunity (Rose, 1987). But *in vivo* production of parasitic antigen as well as requirement of boosting under field condition are limiting factors (Wallach, 2008). It has been documented that Apical membrane antigen 1(AMA1) play an important role at the time of penetration of oocyst into the intestinal wall and used for production of recombinant vaccine (Rugarabamu *et al.*, 2015). Rhomboid like protease involved in processing of MIC also used as recombinant protein and used as DNA or Myco-bacterium bovis vectored formulations (Wang *et al.*, 2014). Glycosyl- phosphatidylinositol (GPI) anchored surface antigen (SAG 1) which is encoded by multi gene families was also tried by various worker as it play important role at the site of attachment to the host prior to invasion (Song *et al.*, 2015). Immune mapped protein 1(IMP1) localized on sporozoite cell membrane also targated as an anticoccidial vaccine candidate for *E. maxima* and *E. tenella* (Yin *et al.*, 2015). Profiline is an actine biding protein and involved in microfilament turnover which play a pivotal role in gliding moment of T. *gondii* (Plattner *et al.*, 2008) and also a ligand for toll like receptors (TLR) 5, 11 and 12 which play important role for initiating the immune response against many intracellular pathogens. During the Eimeria infection up regulation of TLR has also been reported (Zhang *et al.*, 2012).

Role of Cytokine to Augment the Efficacy of Recombinant Vaccine

Type and dose of cytokine affect the efficacy of recombinant antigens like IFNα and lymphotactin were found to improve weight gain during *E. acervulina* challenge, while IL-1β, IL-8, IL-15, IFNy, TGFb4 and lymphotactin decrease parasite replication. It has been reported that administration of IL-2 with Eimeria antigen such as TA4 (Song *et al.*, 2009;) SO7 (Song *et al.*, 2013) and *E. acervulina* antigen zSZ-2 (Shah *et al.*, 2011) can enhance the DNA vaccine induced protective immune response by combination of CD40 ligand with IMP1, enhance antigen presenting cell activation and downstream T-cell mediated effectors functions could be achieved (Yin *et al.*, 2015).

Genetic and Antigenic Diversity of *Eimeria* – Major Hurdle for Development of Coccidial Vaccine

As per the available literature, there are lot of difference between different strain of *Eimeria* for example isolate of *E. tenella* from north India has low haploid while isolates from south India has high haploid (Blake *et al.*, 2015). It might be due to different managemental practices at both geographical regions. Likewise operational taxonomic unit *viz.* OTUxOTUy and OTUz have been reported worldwide (Godwin and Morgan 2015) with the help of advanced technology like Restriction fragment length polymorphism (PCR-RFLP), Amplified fragment length polymorphism (AFLP) and Random amplified polymorphic DNA (RAPD) and variable number tandem repeat (VNTR) (Pegg *et al.*, 2016). Similarly Apical membrane antigen gene are polymorphic and hence create major hindrance in development of recombinant vaccine using the AMA1.

Conclusion

Emergences of new variants among *Eimeria* parasite in different geographical area pose a major menace for the development of vaccine against poultry coccidiosis. Therefore it is imperative to study the evolutionary diversity among gene along with other associate factor for development of effective vaccine.

References

1. Arnott, A., Wapling, J., Mueller, I., Ramsland, P. A., Siba, P. M., Reeder, J. C., and Barry, A. E. (2014). Distinct patterns of diversity, population structure and evolution in the AMA1 genes of sympatric Plasmodium falciparum and Plasmodium vivax populations of Papua New Guinea from an area of similarly high transmission. *Malaria journal, 13*(1), 1-16.
2. Blake, D. P., and Tomley, F. M. (2014). Securing poultry production from the ever-present Eimeria challenge. *Trends in parasitology, 30*(1), 12-19.
3. Chapman, H. D. (1997). Biochemical, genetic and applied aspects of drug resistance in Eimeria parasites of the fowl. *Avian pathology, 26*(2), 221-244.
4. Chapman, H. D., and Jeffers, T. K. (2015). Restoration of sensitivity to salinomycin in Eimeria following 5 flocks of broiler chickens reared in floor-pens using drug programs and vaccination to control coccidiosis *Poultry science, 94*(5), 943-946.
5. Chapman, H.D. (2014). Milestones in avian coccidiosis research: a review. *Poultry science, 93*(3), 501-511.
6. Eckford, S., Grace, K., Harris, C., Reeves, H., Teale, C., Tallentire, C., 2014. UK-VARSS 2013. In: Borriello, S. (Ed.), UK Veterinary Antibiotic Resistance and Sales Surveillance Report.
7. Gedik, Y., Ýz, S. G., Can, H., Döþkaya, A. D., Gürhan, S. Ý. D., Gürüz, Y., and Dö°kaya, M. (2016). Immunogenic multistage recombinant protein vaccine confers partial protection against experimental toxoplasmosis mimicking natural infection in murine model. *Trials in Vaccinology, 5*, 15-23.
8. Godwin, R. M., and Morgan, J. A. (2015). A molecular survey of Eimeria in chickens across Australia. *Veterinary parasitology, 214*(1-2), 16-21.
9. Lassen, B., and Østergaard, S. (2012). Estimation of the economical effects of Eimeria infections in Estonian dairy herds using a stochastic model. *Preventive Veterinary Medicine, 106*(3-4), 258-265.
10. McDonald, V., and Shirley, M. W. (2009). Past and future: vaccination against Eimeria. *Parasitology, 136*(12), 1477-1489.
11. Naciri, M., Chausse, A. M., Fort, G., Bernardet, N., Nerat, F., and de Gussem, K. (2004, June). Utility of anticoccidial sensitivity tests (ASTs) in the prevention of chicken coccidiosis. In 22. World's Poultry Congress. WPSA.
12. Pegg, E., Doyle, K., Clark, E. L., Jatau, I. D., Tomley, F. M., and Blake, D. P. (2016). Application of a new PCR-RFLP panel suggests a restricted

population structure for Eimeria tenella in UK and Irish chickens. *Veterinary parasitology, 229*, 60-67.

13. Plattner, F., Yarovinsky, F., Romero, S., Didry, D., Carlier, M. F., Sher, A., and Soldati-Favre, D. (2008). Toxoplasma profilin is essential for host cell invasion and TLR11-dependent induction of an interleukin-12 response. *Cell host and microbe, 3*(2), 77-87.
14. Reid, A. J., Blake, D. P., Ansari, H. R., Billington, K., Browne, H. P., Bryant, J., . and Pain, A. (2014). Genomic analysis of the causative agents of coccidiosis in domestic chickens. *Genome research, 24*(10), 1676-1685.
15. Rose, M. E. (1987). Immunity to Eimeria infections. *Veterinary immunology and immunopathology, 17*(1-4), 333-343.
16. Rugarabamu, G., Marq, J. B., Guérin, A., Lebrun, M., and SoldatiFavre, D. (2015). Distinct contribution of T oxoplasma gondii rhomboid proteases 4 and 5 to micronemal protein protease 1 activity during invasion. *Molecular microbiology, 97*(2), 244-262.
17. Shah, M. A. A., Song, X., Xu, L., Yan, R., and Li, X. (2011). Construction of DNA vaccines encoding Eimeria acervulina cSZ-2 with chicken IL-2 and IFN-γ and their efficacy against poultry coccidiosis. *Research in veterinary science, 90*(1), 72-77.
18. Shirley, M. W., Smith, A. L., and Tomley, F. M. (2005). The biology of avian Eimeria with an emphasis on their control by vaccination. *Advances in parasitology, 60*, 285-330.
19. Smith, N. C., Wallach, M., Petracca, M., Braun, R., and Eckert, J. (1994). Maternal transfer of antibodies induced by infection with Eimeria maxima partially protects chickens against challenge with Eimeria tenella. *Parasitology, 109*(5), 551-558.
20. Song, H., Qiu, B., Yan, R., Xu, L., Song, X., and Li, X. (2013). The protective efficacy of chimeric SO7/IL-2 DNA vaccine against coccidiosis in chickens. *Research in veterinary science, 94*(3), 562-567.
21. Song, X., Gao, Y., Xu, L., Yan, R., and Li, X. (2015). Partial protection against four species of chicken coccidia induced by multivalent subunit vaccine. *Veterinary parasitology, 212*(3-4), 80-85.
22. Song, X., Xu, L., Yan, R., Huang, X., Shah, M. A. A., and Li, X. (2009). The optimal immunization procedure of DNA vaccine pcDNA–TA4–IL-2 of Eimeria tenella and its cross-immunity to Eimeria necatrix and Eimeria acervulina. *Veterinary parasitology, 159*(1), 30-36.
23. Takala, S. L., and Plowe, C. V. (2009). Genetic diversity and malaria vaccine design, testing and efficacy: preventing and overcoming 'vaccine resistant malaria'. *Parasite immunology, 31*(9), 560-573.

24. Vrba, V., and Pakandl, M. (2015). Host specificity of turkey and chicken Eimeria: controlled cross-transmission studies and a phylogenetic view. *Veterinary parasitology, 208*(3-4), 118-124.

25. Wallach, M. G., Ashash, U., Michael, A., and Smith, N. C. (2008). Field application of a subunit vaccine against an enteric protozoan disease. *PLoS One, 3*(12), e3948.

26. Wang, Q., Chen, L., Li, J., Zheng, J., Cai, N., Gong, P., . and Zhang, X. (2014). A novel recombinant BCG vaccine encoding Eimeria tenella rhomboid and chicken IL-2 induces protective immunity against coccidiosis. *The Korean journal of parasitology, 52*(3), 251.

27. Williams, R. B. (1999). A compartmentalised model for the estimation of the cost of coccidiosis to the world's chicken production industry. *International journal for parasitology, 29*(8), 1209-1229.

28. Williams, R. B., Marshall, R. N., Pagès, M., Dardi, M., and del Cacho, E. (2009). Pathogenesis of Eimeria praecox in chickens: virulence of field strains compared with laboratory strains of E. praecox and Eimeria acervulina. *Avian Pathology, 38*(5), 359-366.

29. Yin, G., Lin, Q., Qiu, J., Qin, M., Tang, X., Suo, X., . and Liu, X. (2015). Immunogenicity and protective efficacy of an Eimeria vaccine candidate based on Eimeria tenella immune mapped protein 1 and chicken CD40 ligand. *Veterinary parasitology, 210*(1-2), 19-24.

30. Zhang, L., Liu, R., Ma, L., Wang, Y., Pan, B., Cai, J., and Wang, M. (2012). Eimeria tenella: expression profiling of toll-like receptors and associated cytokines in the cecum of infected day-old and three-week old SPF chickens. *Experimental parasitology, 130*(4), 442-448.

Chapter 12

African Swine Fever in India: Concept and Update

T.K. Dutta

Department of Veterinary Microbiology,
College of Veterinary Sciences and Animal Husbandry,
Central Agricultural University, Selesih, Aizawl – 796 014, Mizoram

Introduction

African Swine Fever (ASF) is an acute contagious vector borne disease of pigs caused by ASF virus (ASFV) under genus *Asfivirus* and family *Asfarviridae*. It is the only known vector transmitted DNA virus (arbovirus). ASFV is an enveloped and icosahedral DNA virus containing a single molecule of dsDNA as genome with genome length of 170-190 kbp, which encodes about 160-170 genes. The virus is known to be highly resistant to low temperatures. ASFV gets heat inactivated at 56°C for 70 minutes or at 60°C for 20 minutes. In addition, it is inactivated by pH <3.9 or >11.5 in serum-free medium. Serum increases the resistance of the virus, *e.g.* at pH 13.4 – resistance lasts up to 21 hours without serum, and 7 days with serum. The virus is susceptible to ether and chloroform. ASFV gets inactivated by 8/1000 sodium hydroxide (30 minutes), hypochlorites – 2.3 per cent chlorine (30 minutes), 3/1000 formalin (30 minutes), 3 per cent ortho-phenylphenol (30 minutes) and iodine compounds. Alarmingly, the virus remains viable for long periods in blood, faeces and tissues; especially infected, uncooked or undercooked pork products. ASFV multiplies in cytoplasm of the host, as they encode all genes necessary for transcription and replication. The virus can also multiply successfully in vectors (*Ornithodoros* spp.).

Epidemiology

The epidemiology of ASF is complex with different epidemiological patterns of infection occurring in Africa and Europe. ASF occurs through transmission cycles involving domestic pigs, wild boars, wild African suids, and soft ticks. African wild swine (warthogs *Phacochoerus aethiopicus*), bush pigs (*Potamochoerus* spp.), giant forest hogs (*Hylochoerusmeinertzhageni*) are usually infected and act as reservoir hosts of ASFV in Africa. Hosts that demonstrate the disease are domestic pigs (*Sus domestica*), European wild boar, and American wild pigs. Ticks of the genus *Ornithodoros* are considered the natural arthropod host and there exists some speculation that ASFV is a virus of arthropods and that mammalian species, such as domestic swine, represent "accidental hosts".

The virus may be available in the blood, tissues, secretions and excretions of sick and dead animals infected by ASFV. Animals recovered from either acute or chronic infections may become persistently infected and can act as potential carrier of the virus, especially in African wild swine and in domestic pigs in enzootic areas. The virus used to persist for long time in soft ticks of the genus *Ornithodoros*.

The spread and transmission of ASF are unique to each continent of Africa, Europe, and Asia. There are three cycles of ASF epidemiology such as sylvatic, tick–pig, and domestic, with addition of more factors,that is, soft *Ornithodoros* spp. ticks, wild African pigs (mainly warthogs), domestic pigs, and pig-derived products such as pork. In the sylvatic cycle, virus is maintained between warthogs and soft ticks, without causing disease in the warthogs. This may be the reason that the virus spilled over to domestic pigs causing disease. In the tick–pig cycle, ticks act as biological reservoir to maintain the virus and infectious agents gets transmitted among domestic pigs. Such type of cycles was seen in parts of sub-Saharan Africa and also during the epidemic on the Iberian Peninsula in the '60s and '70s of the past century. In the domestic cycle, natural reservoirs do not play a role but disease is transmitted in domestic herds from domestic pigs or pig products. The new and fourth cycle of transmission named wild boar-habitat cycle (proposed) reported in some parts of Europe involves Eurasian wild boar, its habitat and their carcasses. This fourth cycle is involving both direct transmissions between wild boar, and indirect transmission through the habitat. Cold and moist climate favors persistency and human movement are the cause of transmission in domestic pigs. Social and economic factors of pig husbandry are also associated with ASF outbreaks in the domestic cycle. It may be hypothesized that the fourth cycle of ASF virus transmission may be attributed to the first outbreak in Assam and Arunachal Pradesh and subsequently domestic cycle might have played in disease transmission between domestic pigs through contaminated pig products/fomites in rest of the parts of North East India.

The wild boar-habitat cycle is having both direct transmission between infected and susceptible wild boar and indirect transmission through carcasses in the area where wild boar habitat. Such areas get contaminated from infected wild boar carcasses, subsequent scavenging by other wild/other pigs, offer possibilities for both low-dose and high-dose infections, depending on carcass decomposition. It is also observed that the virus may be present in the carcass even if it is a single dead pig for long time and continue to get transmitted to the susceptible animals. The interface contact between wild boar habitat and domestic pigs is critical in transmission of virus which might have caused first outbreak of ASF in India (hypothetical). Environmental persistence of the virus is favoured by a cold and moist climate.

So far, except Sikkim and Tripura all the other 6 states of NER of India including Arunachal Pradesh, Assam, Meghalaya, Mizoram, Mizoram and Nagaland is affected with ASF and more than 50,000 pigs were died.

Clinical Signs and Postmortem Findings

Pigs of all the age groups are equally susceptible. The incubation period varies from 4-19 days. Clinical disease occurs in four forms; peracute, acute, subacute, and chronic. In peracute form, affected animals die without any visible clinical symptoms with mortality rate up to 100 per cent. Acute infections are caused by highly virulent strains and are typically characterized by a high fever, anorexia, lethargy, weakness, recumbency, diarrhea and/or constipation, abdominal pain, hemorrhagic signs, respiratory distress, nasal and conjunctival discharge, and abortions in pregnant females. Death often occurs within 7–10 days after the onset of clinical signs. Moderately virulent strains result in subacute infection (often with high mortality in young animals and much lower mortality in older animals), where the clinical signs often include abortion, fever, and transient hemorrhages with death or recovery occurring within 3–4 weeks. Chronic infections are characterized by intermittent or low fever, loss of appetite, depression and in some instances, result in a fatal infection.

Domestic pigs are most infectious during the incubation period and may act as source of infection for two days before the presentation of clinical disease and recovered pigs may also act as source of infection for one month after the disappearance of clinical signs. The lesions in acute ASF are similar to classical swine fever (CSF) or hog cholera and often difficult to differentiate on the basis of post-mortem examination. The most consistent lesions include enlarged, friable and dark-red spleen, enlarged haemorrhagic gastro-hepatic and renal lymph nodes, petechial to ecchymotic haemorrhages in the kidney, oedema of gall bladder, interlobular oedema of lungs, petechial haemorrhages in the urinary bladder, visceral peritoneum and haemorrhages in lymph nodes of abdominal cavity and myocardium. The pigs infected with highly virulent strains of ASFV exhibit localized areas of reddening or cyanosis of the skin on the tips of ears, extremities of tail, legs and/or over the ham region. Moderately virulent strains also show similar lesions but enlarged spleen is not friable and appears normal in colour. In chronic form, the skin lesions are slightly raised and firm hyperaemic areas of 2 cm or more in diameter to areas of necrosis. Swollen joints and generalized lymphadenopathy are common observations. Pulmonary lobules show areas of consolidation with or without pleural adhesions.

Histological lesions in acute form of disease are seen in the antigen-processing cells of the lymphoreticular system, cortical and stromal areas of lymph nodes, red pulp of spleen and Kuffer cells in liver sinusoids. Haemorrhages in the lymph nodes, kidney and heart are the most common findings. One of the most prominent features in chronic form of ASF is hyperplasia of lymphoreticular tissues.

Diagnosis

ASFV infection is very difficult to differentiate from other closely associated diseases of pigs including CSF, PRRS *etc.* either by clinical or post-mortem examination. Therefore, it is important to diagnose the ASFV infection through recommended laboratory diagnostic techniques only. The laboratory diagnostic techniques are based on detection of the causative agent or immune response to agent. The samples to be collected and sent for laboratory diagnosis are blood,serum, and tissues including spleen, lymph nodes, bone marrow, lung, tonsil, and kidney preferably transported on ice.

Detection of Causative Agent

- Virus isolation: Primary leukocyte culture or porcine Bone marrow cells are commonly used for isolation of virus.
- HAD test: Virus isolation in macrophages and HAD is the gold standard for identification of ASFV required for the first detection of ASF in disease-free regions or in the primary outbreak areas. 8DR protein of ASFV is responsible for HAD property wherein virus in macrophages binds to pig erythrocytes causing rosette and some viruses are non-HAD. HAD ASFV are pathogenic and non-HAD ASFV are non-pathogenic or avirulent. HAD test is cumbersome and is practiced in reference laboratories only. A detailed test protocol is available in OIE manual. The HAD phenotypes of ASFV readily show CPE, where care should be taken while interpreting the non-HAD phenotypes of ASFV which also produce CPE but do not hemadsorb.
- Fluorescent antibody test (FAT): Directly can be used on field samples or those inoculated at the laboratory. Alone can serve as only presumptive diagnostic test along with clinical sings and typical lesions of ASFV.
- ELISA for antigen detection: The sensitivity of the assays is lower than polymerase chain reaction (PCR) or HAD. Hence, usually this method is not recommended for confirmatory diagnosis.

- ☆ Conventional PCR: Should be performed using primer sets targeting the most conserved genome parts so as identify and detect a wide range of ASFV isolates of all the known genotypes.
- ☆ Real-time PCR: The technique is highly sensitive and less time-consuming and even can be used on samples which are not fit for virus isolation or antigen detection assays. The OIE recommends to follow quantitative PCR using real-time PCR primers and probes.

Detection of Immune Response against Causative Agent

The serological tests are the most commonly used diagnostic tests because of their simplicity, relatively lowcost, and less specialized equipment's are needed. The serological tests are also more relevant because as on date there are no commercially available vaccines and hence vaccination is not practiced. Which means presence of antibodies against ASFV indicates either disease is ongoing or it affected the herd. The antibodies appear as early as 7-10 days and can persist for several months or even years.

- ☆ ELISA: The indirect ELISA [59] which is suitable for testing both serum and plasma. However, positive samples should be confirmed by indirect FAT(IFAT), indirect immunoperoxidase test (IPT) or immunoblotting. Different types of ELISA (competitive or blocking) commercially available are being used for the detection of antibodies against ASFV infection.
- ☆ IPT and IFAT test protocols are recommended by OIE. IPT and IFAT are a confirmatory test for sera from areas that are free from ASF and are positive in the ELISA, and for sera from endemic areas that give an inconclusive result in the ELISA.
- ☆ Immunoblotting test: The procedure is used as alternate to IFAT and IPT to confirm equivocal results with individual sera. The tests perform better than other tests even in the detection of weak positive samples for ASFV antibodies.

Prevention and Control

Like most of the viral infections, no specific treatment is available against ASFV infection. In addition, to date, no vaccine is also available due to several key factors, including the lack of identification of protective antigens, incomplete understanding of virus-host cell interactions, and inadequate knowledge relative to the diversity of viral strains currently circulating in natural reservoirs. Many vaccine options have been tried with varying levels of success, including vaccines with naturally or experimentally deleted genes, subunit vaccines based on recombinant proteins, and DNA vaccines. Vaccine development for ASFV is ongoing and challenging due to the range of genetic and antigenic variability as well as the myriad of strategies utilized by the virus to evade the host's immune response. Further work is essential to develop a vaccine that is both safe and provides a high degree of protection across virulent ASFV strains. The most appropriate method to control the disease is slaughter and proper carcass disposal of infected as well as suspected pigs followed by disinfection of contaminated area. Also, restricted movements of live pigs, their products from an infected area to non-infected area or vice-versa should be adopted as effective control measure. ASF is a vector-borne disease, so elimination of its vector is an important steps in controlling the disease. Care should also be taken for fomites (clothes, utensils, vehicles, shoes, *etc.*), which play an important role in spreading of infection. Strict biosecurity measures should be adopted in commercial as well as backyard farms to prevent the entry of infection in the non-infected areas. Swill feeding must be regulated by invoking stringent effort. Fences must be installed to block intruders such as roaming wild boar. Strict regulations and controls on food waste from international flights, and ships should be adopted at national level.

Conclusion

African swine fever (ASF) is a highly contagious and severe viral disease affecting domestic and wild pigs and considered as one of the most serious threat to pig husbandry globally. As the outbreak of ASF is reported from North East India, it may pose a serious threat to the entire pig husbandry because of its extremely high level of morbidity and mortality. It is already indicated that there is no specific treatment or protective vaccines are available against the disease. The disease transmission is also very stable through various animate and inanimate objects including ticks as biological and mechanical carrier. Therefore, utmost precautions need to taken by the Governments to protect the pigs and the poor and marginal pig farmers of India to sustain the rural economy. Knowledge gained during the first entry and outbreak of PRRS in pig population in India in general and in the North eastern India in particular in 2011 should be applied in containment and control of the disease. Restriction of movement of animals and animal handlers including vehicles, feeds, processed pork products *etc.* from affected areas should be strictly restricted by implementation of effective rules and regulations. The National Action Plan for containment, prevention and control of African Swine Fever (June, 2020) should be followed in implementation of prevention measures by all the state governments in India.

References

1. Brown C. and Torres A., Eds. (2008). - USAHA Foreign Animal Diseases, Seventh Edition. Committee of Foreign and Emerging Diseases of the US Animal Health Association. Boca Publications Group, Inc.
2. Coetzer J.A.W. and Tustin R.C. Eds. (2004). - Infectious Diseases of Livestock, 2nd Edition. Oxford University Press.

3. Fauquet C., Fauquet M., and Mayo M.A. (2005). - Virus Taxonomy: VIII Report of the International Committee on Taxonomy of Viruses. Academic Press.

4. Kahn C.M., Ed. (2005). - Merck Veterinary Manual. Merck and Co. Inc. and Merial Ltd.

5. Kumar S., Dutta T. K., Roychoudhury P., Das H., Akter F. and Subudhi P.K. (2020). African swine fever: An emerging threat to pig husbandry in India. Indian Journal of Animal Health. 59: 11-16.

6. Patil S. S., Suresh KP, Vashist V, Prajapati A, Pattnaik B and Roy P (2020). African swine fever: A permanent threat to Indian pigs. Veterinary World. 13: 2275-2285.

7. Spickler A.R., and Roth J.A. Iowa State University, College of Veterinary Medicine - http: //www.cfsph.iastate.edu/DiseaseInfo/factsheets.htm

8. World Organisation for Animal Health (2012). - Terrestrial Animal Health Code. OIE, Paris.

9. World Organisation for Animal Health (2012). - Manual of Diagnostic Tests and Vaccines for Terrestrial Animals. OIE, Paris.

Chapter 13

An Overview of Diagnosis of Mycoplasma Infections

Amita Tiwari

Assistant Professor, Department of Veterinary Medicine,
College of Veterinary Science and A.H., NDVSU, Jabalpur, Madhya Pradesh

Mycoplasmas are the smallest free-living organisms which are members of class Mollicutes, order Mycoplasmatales. There are approximately 200 known species of Mollicutes from animals, only a small number, mainly Mycoplasma species, are known as pathogens. They have no cell wall and are surrounded by a membrane. These organisms have global distribution causing serious diseases in animals worldwide including mastitis, arthritis, pneumonia, otitis media and reproductive disorders. Mycoplasmas are highly contagious which cause severe diseases and require rapid and accurate diagnosis to prevent and control disease outbreaks.

The first Mycoplasma species was isolated from cattle in 1898 as *Mycoplasmamycoides* subsp. mycoides and was known as the causative agent of contagious bovine pleuropneumonia. In the 19th century, this disease spread globally and then it was eradicated from most continents, however, it remains in many African countries.

There are several Mycoplasma species in cattle with varying degrees of clinical significance. These include *M. californicum, M. bovigenitalium, M. bovirhinis, M. bovoculi, M. canis, M. canadense, M. alkalescens, M. Arginini*etc. However, *Mycoplasma bovis* is currently known as one of the most important and frequently associated with disease in cattle worldwide. The highly contagious nature of Mycoplasma spp., their poor responsiveness to treatment and associated culling implications for affected stock make rapid and accurate diagnosis important for control and prevention of disease outbreaks. So here we discuss the use of different diagnostic methods available to identify Mycoplasma species.

Culture Techniques

Traditionally, the identification and detection of Mycoplasma species has been performed via microbial culture. In order for mycoplasma infection to be diagnosed by culture, the animal must be shedding viable organisms at the time of sample collection. Intermittent shedding is frequently reported with chronic and subclinical mycoplasma mastitis cases, which can result in a diagnosis failure from a single sample.

The culture media after inoculatioc is incubated for 7–10 days at 37.8°C and 5 per cent CO_2, resulting in the growth of micro-colonies with a morphological appearance of "fried eggs" for most Mycoplasma species, which are visible via light microscopy (Quinn *et al.*, 2013). This appearance is because of the central portion of the colony embedding itself into the agar surrounded by a zone of surface growth.

If we have to diagnose mycoplasma in milk samples, the limit of detection for culture has been reported as low as 272 cfu/mL in milk (Sachse *et al.*, 2010). When mycoplasma is present in low concentrations in milk, to concentrate the cfu/mL and improve recovery by culture, the milk can be centrifuged first and suspended in a small volume of saline before culturing. Identification of mycoplasmas via culture is the traditional method because of its relative simplicity and low cost but it have several limitations. Methods of sample collection, handling and storage are essential to prevent or minimize the overgrowth of other bacteria in the sample and to maximize mycoplasma viability and growth on the diagnostic media. Samples should be collected aseptically into sterile tubes or containers and then kept at 48°C and transported to the laboratory for culturing as soon as possible. If culture cannot be performed within 2 days, milk samples should be frozen. Ideally, frozen samples should be thawed at room temperature to ensure maximum recovery, with repeat freeze thawing causing a further decrease in viable organisms (Biddle *et al.*, 2004).

Mycoplasma recovery rate decreases with increased time. With appropriate sample collection and storage protocols, mycoplasmas still exhibits a slow rate of growth, with colonies not visible for a minimum of 5 days and sometimes requiring up to 10 days. It is problematic because rapid diagnosis of infection is desirable to facilitate removal of infected animals from the herd to minimize transmission.

Serological Techniques

While culture techniques relies upon demonstrating the presence of the mycoplasma organism, indirect ELISAs are available that demonstrate the presence of anti-Mycoplasma antibodies in plasma, serum, and milk samples. Their purpose is to identify animals that have been exposed to the pathogen and had time to mount a humoral immunological response (Parker *et al.*, 2018). Because of M. bovis currently being considered the most common and important species, much of the assay development and research available is specific to this species. This includes the commercially available indirect ELISA kits for anti-M. bovis antibody detection. Recent studies have evaluated the use of the anti-M. bovis antibody ELISA kit on bulk tank milk samples to estimate herd-level prevalence.

When developing an ELISA assay, the potential for cross-reactivity of an *M. bovis* specific ELISA with other bacterial species found in cattle and ruminants is difficult to assess but is important from a diagnostic perspective, particularly if these species are not of clinical importance.

Molecular Techniques

The use of PCR to detect mycoplasma species from various sample types has demonstrated a higher efficiency, specificity, and sensitivity for laboratory diagnosis when compared with conventional culture based diagnostic methodologies. Unlike culture, the organism does not need to be viable for detection. Results are produced within 24 hours, rendering the PCR method more rapid than culture.

A more recent study validated a conventional multiplex PCR followed by sequence analysis to allow the identification and discrimination of several Mycoplasma and *Acholeplasma* spp. As technology has improved, real time PCR has also been developed and utilized in mycoplasma detection. To achieve greater specificity, the fluorescent reporter probe method of real time PCR was developed, typically utilizing a hydrolysis probe. A recent study developed three individual probe-based real-time PCR assays to detect *M. bovis*, *M. californicum* and *M. bovigenitalium* in bovine milk and tissue samples (Boonyayatra*et al.*, 2012).

Since mycoplasma infections are not limited to mastitis, it is important that other sample types such as joint fluid aspirates and mucosal surface swabs are also needed to be investigated. Haemotropic mycoplasmas (or haemoplasmas) are the causative agents of infectious anaemia in many mammalian species. They were previously known as Haemobartonella and Eperythrozoon species. PCR is the method of choice to diagnose and differentiate haemoplasma infections. Specific conventional and quantitative real-time PCR systems have been introduced and are now considered the gold standard for the detection and differentiation of feline and canine haemoplasma species (Willi *et al.*, 2010).

Conclusion

Mycoplasmas can cause serious disease in animals resulting in significant negative economic and welfare impacts. Rapid diagnosis facilitates prompt disease control and prevention. While culture and PCR rely on the animal shedding the organism at the time of sampling, an ELISA can measure past exposure to the organism in the form of a humoral antibody response which is usually sustained for a period of time after infection and therefore does not require detection of an organism that may be shed intermittently or is no longer present. So, all the diagnostic methods have both advantages and disadvantages. However, their combined use will complement each other based on their individual strengths.

References

1. Biddle, M.K., Fox, L.K., Hancock, D.D., Gaskins, C.T., Evans, M.A. (2004). Effects of storage time and thawing methods on the recovery of Mycoplasma species in milk samples from cows with intramammary infections. J Dairy Sci., 87: 933–936.

2. Boonyayatra, S., Fox, L.K., Besser, T.E., Sawant, A., Gay, J.M., Raviv, Z. (2012).A PCR assay and PCR-restriction fragment length polymorphism combination identifying the 3 primary Mycoplasma species causing mastitis. J Dairy Sci., 95: 196–205.

3. Parker, A.M., Sheehy, P.A., Hazelton, M. S., Bosward, K. L., House, J.K. (2018). A review of mycoplasma diagnostics in cattle. wileyonlinelibrary.com/journal/jvim, 1–12.

4. Quinn, P.J., Markey, B.K., Leonard, F.C., Fanning, S., Maguire, D. (2013). Veterinary Microbiology and Microbial Disease. 2nd ed. Hoboken: Wiley.

5. Sachse, K., Salam, H.S.H., Diller, R., Schubert, E., Hoffmann, B., Hotzel, H. (2010). Use of a novel real-time PCR technique to monitor and quantitate *Mycoplasma bovis* infection in cattle herds with mastitis and respiratory disease. Vet J., 186: 299–303.

6. Willi, B., Novacco, M., Meli, M. L., Wolf-Jäckel, G. A., Boretti, F. S., Wengi, N., Lutz, H., Hofmann-Lehmann, R. (2010). Haemotropic mycoplasmas of cats and dogs: transmission, diagnosis, prevalence and importance in Europe. Schweiz. Arch. Tierheilk, 237 – 244.

Chapter 14

One Health: A Holistic Approach to Restrict Outbreaks

Namita Joshi and R..K. Joshi

College of Veterinary Science and Animal Husbandry,
Acharya Narenrdra Deva Krishi Evam Prodyogik Vishwavidyalaya,
Kumarganj – 224 229, Ayodhya, Uttar Pradesh

It has been increasingly clear that majority of infectious diseases, which have emerged over the past three decades, have originated from animals, especially wildlife (Mackenzie and Jeggo,2019). These, newly emerging and re-emerging infections are now recognized as a global problem, and 75 per cent of these are potentially zoonotic. It is reported that 61 per cent of known pathogens can infect multiple animal species thus increasing their potential of survival and transmission in the ecosystem (Woolhouse and Gowtage-Sequeria, 2005, Jones, 2008). Emerging infectious disease events are not simply about pathogens jumping species barriers, they are made of complex relations that involve socioeconomic and socio-political drivers and their consequences extend beyond the impact of the disease itself.

Therefore, a collaborative and multi-disciplinary approach is essentially needed to understand the ecology of each emerging zoonotic disease in order to undertake a risk assessment, and to develop plans for response and control. One Health is the only approach where multiple disciplines work under one umbrella. This approach conceptualizes that the health of people is closely connected to the health of animals and their environment. The purpose of this article is to bring the attention of healthcare professionals, researchers and policy makers toward the One Health paradigm for the betterment of public health and to prepare ourselves for combating future epidemic or pandemic.

What is One Health?

According to WHO, health is a "state of complete physical, mental and social well-being and not merely the absence of disease and infirmity". To attain a healthy state, humans cannot overlook the essential contribution of coexisting species—animals and plants—since we share the same environment and ecosystem. Human health, animal health, and environmental factors are inextricably dependent on each other. Therefore, "Health" as a concept should not be confined to the well-being of human beings only rather it should include health of all. One health is defined by the One Health Commission as "the collaborative effort of multiple disciplines to obtain optimal health for people, animals, and our environment." In another definition, the One Health Initiative Task Force (OHITF) defines one health as "the promotion, improvement and defence for the health and well-being of all species by enhancing cooperation and collaboration between physicians, veterinarians, and other scientific health professionals and by promoting strengths in leadership and management to achieve these goals." The concept calls for application of a coordinated, collaborative, multidisciplinary and cross-sectoral approach to address potential or existing risks that originate at the animal-human-ecosystems interface.

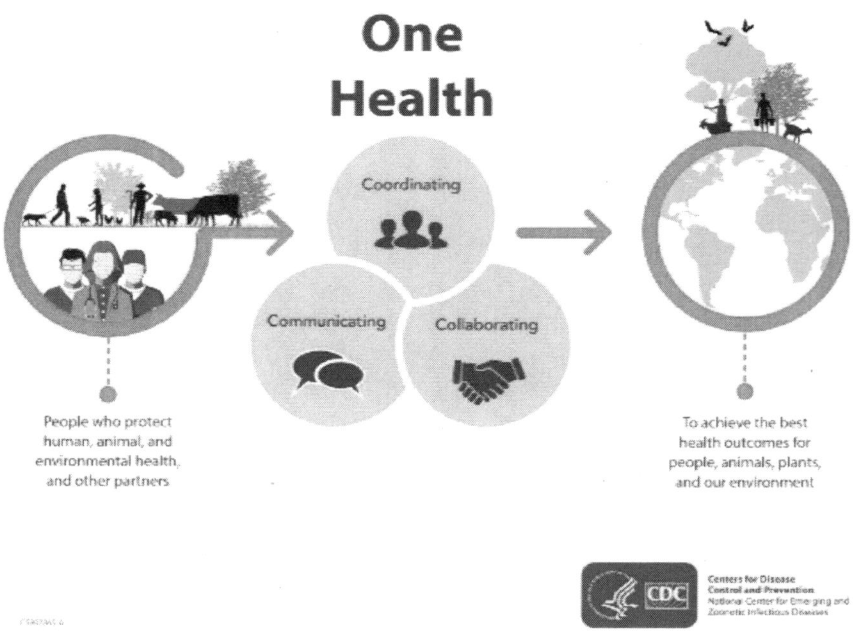

History

The fascinating history of human and animal medicine reveals that the concept of one health is not new, rather it has evolved over a period of time. The scientific knowledge of health and disease developed due to the keen observations of Greek and Roman scholars as evidenced from their writing. Hippocrates, *'Father of Medicine'*, in 5th century pointed out the association between man's environment and his state of health. Aristole indicated the knowledge of such relationship in rabies and plague. In ancient time, there was no clear-cut demarcation between human and animal medicine. It was customary for a person to practice both the professions simultaneously. During middle age, services provided by veterinarians and human doctors became somewhat distinct; although veterinary doctors were still participating in some areas of human medicine like orthopaedics and obstetrics. For example, in 1500 JocobNufer, a veterinarian performed first successful caesarean operation on a woman. Dr. Otto Stader, a small animal veterinarian developed the first form of external skeleton fixation *i.e.* the splint to stabilize fracture in dog. During World War II, same procedure was used for fracture management by Navy surgeon in sailors. This one medicine concept was lost over a period of time despite numerous discoveries and human and veterinary discipline developed as separate profession. The greatest philosopher, German Physician and *'Father of cellular pathology'* "Rudolf Virchow" (1821-1902) regenerated the concept of one health. He stated that between human and animal medicine, there is no dividing line nor there should be. The object is different but the experience obtained constitutes the basis of all medicine. The term **One Medicine** was first coined by Calwin W. Schwabe (1927-2006) in 1984 in his text book "Veterinary medicine and human health" to bring renewed interest to synergy that can emerge when human and veterinary doctors collaborate. This One Medicine concept later evolved into "One World, One Health" when an international conference was held by wildlife conservation society in 2004 and eventually resulted into modified version *i.e.* One Health following the pandemic of Avian Influenza.

Factors Fostering the Need of One Health

One Health is not new, but it has become more important in recent years. This is because many factors have changed interactions between people, animals, plants, and our environment.

- Population explosion: Human population is growing and expanding into new geographic areas. As a result, more people are forced to live in close contact with wild and domestic animals, both livestock and pets.
- Erratic human life style: Animals play an important role in our lives, whether for food, fiber, livelihoods, travel, sport, education, or companionship. Close contact with animals and their environments provides more opportunities for diseases to pass between animals and human.
- Climate change: The earth has experienced changes in climate due to industrialization, urbanization and automobiles expansion. The

toxic fumes and chemical pollutants have resulted in global warming leading to increase in vector population.

- Loss of Biodiversity: Disruptions in environmental conditions and habitats for unsustainable development has caused loss of biodiversity and any disturbance ecosystem can provide new opportunities for diseases to pass from animals to man.
- Change in land use pattern and intensive farming practices has caused diseases like JE and bird flu *etc.*
- Mutations and variations in pathogens
- International trade and travel: The movement of people, animals, and animal products has increased in 21st century especially illegal trade of wildlife. As a result, diseases can spread quickly across borders and around the globe.
- Our persistent destruction of wildlife due to explosive population growth has brought us near to the wild animals and plants that foster pathogens that can move from animals to humans.

These changes have led to the spread of existing or known (endemic) and new or emerging zoonotic diseases.

One Health in 21st Century

The term 'One Health' was first used in 2003–2004, following the emergence of severe acute respiratory disease (SARS) in early 2003 and subsequently after the spread of highly pathogenic avian influenza H5N1, the concept got momentum in scientific community. **Manhattan Principles'** which were derived at the meeting of the "Wildlife Conservation Society" in 2004, were a vital step towards recognising the critical importance of collaborative and cross-disciplinary approach for responding to emerging and resurging diseases. These principles gave a thrust for idea of inclusion of wildlife health as an essential component of global disease prevention, surveillance, control, and mitigation (Mackenzie and Jeggo, 2019). Subsequently, the tripartite partners (WHO, FAO and OIE) took the cognizance of one health concept and added antimicrobial resistance, rabies and zoonotic influenza as main global problems. In October 2017, the tripartite partners made a commitment to provide multi-sectoral and collaborative leadership in addressing health challenges. The strategic framework was expected to focus on:

- The strengthening and modernisation of early warning and surveillance/monitoring systems
- The foresight, preparedness and response to emerging, re-emerging and neglected infectious diseases
- The encouragement and the promotion of coordinated research and development to achieve a common understanding of the highest priority zoonotic diseases

- The challenge that represents food safety requiring a multi-sector approach in the context of reinforcing food security

One Health Efforts at International Level

Establishment of division of VPH at CDC in 1947 was first step towards one health by James H Steel, a renowned veterinarian who contributed to the understanding of how the diseases spread between human and animals. Since the end of 2005, there had been increasing interest in one health initiatives and numerous international meetings and symposia had been held across the world at various places like Winnipeg (Manitoba, Canada, March 2009), Hanoi (Vietnam, April 2010), Stone Mountain (Georgia, US, May 2010) and Melbourne (Australia, February 2011).

OIE has modernised its worldwide information system on animal diseases (including zoonoses) with the creation of world animal health information system (WAHIS), a mechanism whereby all countries are linked on-line to a central server that collects all the compulsory notifications sent to the OIE, covering 100 priority, terrestrial and aquatic animal diseases.

WHO has adopted the International Health Regulations, placing new obligations for its Members. The OIE, WHO and FAO have created the Global Early Warning System (GLEWS), a platform shared by the three organisations to improve early warning on animal diseases and zoonoses worldwide.

FAO: In 2011, the FAO put forth a strategic action plan for One Health, which had the objective to strengthen food security by improving animal production and veterinary services. The plan called for action in improving collaboration between animal, human, and environmental health sectors.

American Veterinary Medical Association: established one health initiative task force in 2006.

American Medical Association (2007): passed one health resolution to promote partnership between multiple discipline.

CDC established one health office in 2009: This establishment undertook **one health** activities in 15 countries including India and actively involved in prioritization of one health zoonoses. Diseases most commonly prioritized worldwide include rabies, brucellosis, influenza, Ebola virus and rift valley fever.

WHO, FAO and OIE came together with UNICF and WHO to develop a framework entitled "Contributing to One World, One Health-A Strategic Framework for Reducing Risks of Infectious Diseases at the Animal-Human-Ecosystems Interface" in 2008.

One Health Commission: It was officially chartered in Washington D.C. in 2009, with Roger Mahr as the founding CEO. Currently, this commission is chaired by a veterinarian, Dr Cheryl Stroud since 2013.

Journals: The field of One Health has expanded substantially since 2000. A number of publications and journals came up on topic like One Health, EcoHealth, The Lancet Planetary Health, The One Health Platform, The International Association of Ecology and Health, The Planetary Health Alliance *etc.* (Cunnigham, 2017). This shows the growing internet of scientific community towards "One Health".

One Health Day: In 2016, The One Health Commission, One Health Platform, and One Health Initiative Team deemed International One Health Day to be observed on 3rd November (*www.onehealthcommission.org.*)

One Health initiatives for AMR: The tripartite partners have recognised antimicrobial resistance (AMR) as major public health threat and recommended it to be part of one health approach.

Global Action Plan – 2015: This planaimed to ensure prevention and treatment of infectious diseases with safe and effective medication.

The Global Antimicrobial Resistance Surveillance System (GLASS): It supports a standardized approach to the collection, analysis and sharing of data to provide information for decision-making and to drive local, national and regional action plans.

Global Antibiotic Research and Development Partnership (GARDP): GARDP, created by WHO and Drug and Neglected Diseases initiative (DNDi) for developing new treatments for drug resistance infections that pose the greatest threat to health. It mobilizes the partner to develop new and improved antibiotics.

Global Alliance for Rabies Control: It is a mission to prevent human deaths from dog mediated rabies and relieve burden of rabies in other animal population, especially dogs. GARC aims to eliminate human deaths from dog rabies by 2030.

The One Health Global Network: It aims at facilitating coordination and providing linkages; offering a global geographic dimension and optimal complementarity between initiatives.

One Health Initiatives in India:A stellar example of One Health being operationalized in field was seen in late 1950s. It helped to discover source of Kyasanur Forest disease, a highly dangerous haemorrhagic fever and more threatening than COVID 19. It was locally named as monkey fever because of link between monkey deaths and human infections in Shimoga district of Karnataka where it emerged in 1957. It brought together diverse entities like Rockefeller Foundation (provided financial and technical support) and National Institute of Virology (committed team of researchers who combed the forest), WHO (supported birdman of India, Salim Ali for tagging to rule the possibilities that they were carrying pathogens and Bombay Natural History Society. In 2018, Kerala reacted quickly and efficiently to tackle Nipah Virus outbreak and successfully managed to restrict it to 23 cases. This success is credited to strong political will to seek help from multidisciplinary team of national and international experts. The Government of India has taken some initiatives to tackle burgeoning problems such as antimicrobial resistance, zoonotic diseases, and food safety using the One Health approach, but it is still in embryonic stage. India has a long way to go in terms of human and material resources to implement One Health approach.

- Public Health Foundation of India
- National Standing Committee on Zoonoses (2007)
- Government initiated control program for zoonotic and highly communicable diseases such as NADCP (2019) for brucellosis and FMD, National Rabies Control Programme, National tuberculosis control programme.
- National Institute of Zoonoses (NIZ): established on 6th July 2010 at Nagpur by the National Association of Welfare of Animals and Research in collaboration with Indian Medical Association (Nagpur Chapter). In this context, two collaborative research projects, funded by the ICAR and ICMR were initiated to conduct the epidemiological surveillance of selected zoonotic diseases in Central India.
- National Institute of One health: Since the inception of NIZ, discussions were made for setting up of a dedicated facility and finally ICMR decided to establish NIOH at Nagpur in 2019. This will work as satellite centre under National Institute of Virology Pune and MAFSU, Nagpur and help the country for better preparedness against future outbreak.
- One Health Network South Asia:
- To enhance the capacity in epidemiology and bio-security in southasian region.
- Centre for One health education advocacy research and training (COHERT): Feb 2014 in Kerala Veterinary and Animal Science University
- Launched two post graduate courses on one health
- Centre provide Post Graduate degree in one health through distance learning.

One Health Alliance of South Asia aims to predict and prevent emerging infectious diseases on the Indian subcontinent.

One Health Network South Asia: It is proposed that the One Health Hub in India would be connected to similar One Health Hubs throughout South Asia, as part of the regional South Asia One Health Network, which supports One Health communication, collaboration, professional networking, projects and training activities across the South Asia region.

Tripartite Line of Action

In One Health expert meeting held at the Stone Mountain (May, 2010), participants agreed that One Health should be promoted by various institutions but not "possessed" or "mastered" by any institution (Rubin, C., 2011).WHO, FAO and OIE experts recommended concrete line of actions to move the concept from vision to implementation by:

- Cataloguing and developing One Health trainings and curricula.
- Establishing a One Health global network among various sectors. Country-level needs assessment.
- Building capacity at the country level.
- Developing a business case to promote donor support.
- Gathering evidence for proof of concept through literature reviews and prospective studies.

Wildlife and One Health

Wildlife generally helps us to stay healthy in a healthy ecosystem by harbouring various pathogens. Pathogens keep circulating in milder forms in wild animals living in undistributed habitats but if the balance is disturbed; they move out of their niche and come in contact with human population. Sometimes, they jump the interspecies barrier by undergoing mutation. Because of this, they can sometimes serve as early warning signs of potential human illness. For example, birds often start dying well before people in the same area get sick with West Nile virus infection. Death of monkeys in Kysanur forest is indicative of KFD. If we throw a glance over the emerging infectious diseases that have struck human being in last few decades, we find that they all have connection with wildlife. One of the reasons for the Ebola outbreak in West Africa was deforestation bridging the gap between wildlife and human habitat. The persistent destruction of wildlife due to forest fire, climate change, hunting for illegal trade and research purposes has provided opportunity to pathogens to jump from animals to humans. Furthermore, unsustainable development across the world has caused massive loss of biodiversity posing a threat to human and animal health. Thus, emerging infectious diseases can be both a cause or consequence of biodiversity loss. (Daszak *et al.*, 2000). The importance of wildlife as reservoirs of parasitic zoonoses has also been widely recognized *e.g.* American and African trypanosomiasis, leishmaniasis, giardiasis, cryptosporidiosis *etc.* The increasing encroachment of people and livestock into wildlife habitats provided a multifaceted need to study emerging infectious diseases at the human-wildlife interface (Wood *et al*, 2012).

Constraints

In India, range of permission is required to do research on One Health based Problems. The regulatory framework for doing one health research in India in collaboration with international agency require permission from multiple authorities such as ICMR, Ministries of External Affairs and Finance, Directorate General of Armed Forces, National Biodiversity Authority, Institutional Animal Ethical Committee and State Health authority. Additional permission is required from state forest department, biodiversity board for accessing biological resource within natural landscape. Although necessity for research permits is not being questioned but the long waiting periods to seek range of permission raises issues whether we are unwittingly hampering our ability to rapidly respond to emerging threats from infectious diseases. The major bottlenecks in implementing OH include absence of a legal framework to implement OH, poor coordination among different governmental and private agencies, lack of proper surveillance of animal diseases, poor data-sharing mechanism across sectors, and limited budget. Implementing systematic zoonotic surveillance; regulated antibiotic use among humans and animals; development of a zoonotic registry in the country; constitution of a wide network of academic, research, pharmaceutical, and various implementation stakeholders from different sectors is the need of the hour to effectively use OH in order to combat increasing zoonotic diseases.

Conclusion

One Health is a well-known concept since centuries; however, it has been at the fringe of most health policies rather than being the central theme. Although, all stakeholders of health system and policy maker have agreed on this theory, the transition from a vision to practical application seems to be inconspicuous. COVID-19 pandemic has caused massive damage to the world economy and continues to peril human lives everywhere. The social distancing, lockdown, and covid appropriate behaviour may be short-term preventive measures but are insufficient and difficult to sustain in the long run. We have seen that ignorance of the principles of One Health approach in the current pandemic has collapsed our health care system and our health policy has proved to be the Achilles heel. Thus, it is long overdue that we change our unidimensional approach regarding the control and prevention of outbreaks of infectious diseases. A rational practice of the One Health strategy should be our utmost priority to control the ongoing grave situation.

References

1. Daszak P, Cunningham AA, Hyatt AD. (2000). Emerging infectious diseases of wildlife— threats to biodiversity and human health. Science 287, 443–449.

2. Jones, K. E., Patel, N. G., Levy, M. A, *et al.* (2008). Global trends in emerging infectious diseases. *Nature,* **451:** 990–93.

3. Jorwal, Pooja, Bharadwaj, Swati and Jorwal, Pankaj (2020) One Health Approach and Covid 19: A perspective. Journal of Family Medicine and Primary Care. V9(12) PMC7928117.

4. Mackenzie, John S. and Jeggo, Martin (2019) The one health approach: why is it so important? *Trop. Med. Infect. Dis.* 4(2)88.

5. Rubin, C (2011) Operationalizing One Health: The Stone Mountain Meeting. 1st International One Health Congress, February 16, 2011. www.cdc.gov/onehealth.

6. Schwabe, Calwin W. (1984) Veterinary Interests in Public Health in a textbook on Veterinary Medicine and Human Health. Williams and Wilkins Publication 3rd revised edition. pp 34-60. ISBN-13: 978-0683075946

7. Woolhouse, M. E. J. and Gowtage-Sequeria, S. (2005) Host range and emerging and re-emerging pathogens. Emerg. Infect. Dis. 11, 1842–1847.

8. Wood J. L. N., Melissa Leach and Cunningham, A. A. (2012). A framework for the study of zoonotic disease emergence and its drivers: spillover of bat pathogens as a case study. Philosophical Transaction B. The Royal. Society. B 367, 2881–2892.

Chapter 15

Current Trends in Diagnosis and Prophylaxis of Canine Morbillivirus (Canine Distemper Virus)

K.K. Rajak[1], M. Bhatt, Kiran[1], Monu Karki[1], Vishal Rai[1], A. Fayaz[1], A. Kumar[1], A.K. Yadav[1], V. Upamanyu[2] and R.P. Singh[3]*

[1]*Division of Biological Products,*
[2]*Division of Biological Standardization,*
ICAR–Indian Veterinary Research Institute, Izatnagar – 263 122, Uttar Pradesh
[3]*ICAR-Directorate of Foot and Mouth Disease, Mukteswar – 263 138, Uttar Pradesh*
**e-mail: kaushalvirol@gmail.com*

Introduction

Canine distemper (CD), caused by *Canine morbillivirus*, is a highly contagious and often fatal disease of dogs and other carnivores worldwide. The domestic dogs are the main reservoir of *Canine morbillivirus* having fatality rate second highest after rabies (Deem *et al.*, 2000). *Canine morbillivirus*, also known as canine distemper virus (CDV) is an enveloped, single-stranded, negative-sense, non-segmented RNA virus of the genus *Morbillivirus* within the family *Paramyxoviridae* and the order *Mononegavirales* (Pope *et al.*, 2016).CDV is highly susceptible to ultraviolet light, heat and drying conditions and is easily destroyed at temperatures greater than 50° to 60°C within 30 minutes. Various chemical agents like chloroform, ether (<0.5 per cent), formalin, phenol (0.75 per cent), and quaternary ammonium disinfectants (0.3 per cent) also destroy the virus. The virus can survive at freezing temperature of -65°C for at least 7 years (Greene and Appel, 2006). The genome encodes one non-structural protein (C) and six structural proteins: Large protein (L), Haemagglutinin (H), Phosphoprotein (P), Nucleocapsid protein (N), Fusion protein (F) and Matrix protein (M) in the order 3′-N-P-M-F-H-L-5′. The non-structural protein (C) is produced by an alternative open reading frame in the P gene (Lamb and Kolakofsky, 2001).

CDV is monotypic in nature but classified into 18 different genetic lineages (H-gene based) circulating across the world. A new genetic lineage found to circulate in India *i.e.* India-1/Asia-5 has been reported recently (Bhatt *et al.*, 2019). Young, unvaccinated puppies and non-immunized older dogs tend to be more susceptible to the disease. Death among puppies may often reach up to 80 per cent. Domesticated dogs are the main reservoirs of CDV and are responsible for introducing the disease to previously unexposed wildlife or vice-versa (Viana *et al.*, 2016). The virus has a wide host range and the ability of virus to switch hosts causes a serious conservation threat to many species of carnivores. Although modified live and recombinant vectored vaccines are available to protect the animals from the disease, outbreaks in vaccinated dog population and wildlife are alarming. Recent outbreaks in wildlife and dog population indicate that all the control and preventive measures are hard-pressed or failed when the virus is on its march. In India, disease has disseminated to wildlife sanctuaries where huge mortality has been noticed in Lions and leopard (https://www.hindustantimes.com/india-news/lion-death-toll-in-gir-rises-to-21-officials-admit-some-succumbed-to-viral-infection/story-JBq6AVb4wvZqbbPj YVnZVN.html). Of late, mortality of 23 Asiatic lion in Gir forest of Gujarat due to CDV has raised the severe concern to contain the disease in India (Bhatt *et al.*, 2019).

Host Range and Transmission

Although CDV was initially described as an infectious disease of domestic dogs, its ability to infect multiple species has led to mass mortalities in a range of carnivore species from wild canids to felids, hyaenids, procyonids, ailurids, ursids, mustelids and viverrids. Distemper outbreaks have also been reported in marine mammals, including Baikal and Caspian seals (Mamaev*et al.*, 1995; Kennedy *et al.*, 2000). Recently, CDV was reported in non-human primates [rhesus monkey (*Macacamulatta*) and cynomolgus macaques (*Macacafascicularis*)] with high mortality rates (Loots *et al.*, 2017).

Canine distemper is a highly contagious and is readily transmitted between susceptible hosts through contact or aerosolized oral, respiratory and ocular fluids and exudates containing the pathogen (Kapil *et al.*, 2011). During the acute phase of infection, other body excretions and secretions (e. g. urine, faeces, skin) can also contain the virus. Viral shedding may follow the infection for up to 90 days and occurs even if the animal was sub-clinically infected. Transmission of CDV is largely dependent on the close association between affected and susceptible animals. Owing to their wide distribution, domestic dogs are key reservoirs for a variety of diseases and are considered the primary reservoir for CDV infection. Domestic dogs, from communities surrounding protected wildlife areas, are often unvaccinated and come in contact with wildlife, as both may wander several kilometers in and out of the protected areas, increasing the risk of disease transmission, especially if these areas are unfenced. This risk of disease transmission between domestic dogs and wildlife is further augmented by a general lack of vaccination programs, particularly in rural areas. Interactions among potential vectors of CDV, such as jackal, hyenas and lions provide a potential mechanism for subsequent cross-species transmission (Cleavelandet *al.*, 2000).

Pathogenesis

Two major host cellular receptors have been identified that play a critical role in CDV pathogenesis: the signaling lymphocyte activation molecule (SLAM, CD150) and nectin-4 (poliovirus-receptor-like-4) (Tatsuo *et al.*, 2001; Mühlebach*et al.*, 2011; Pratakpiriya *et al.*, 2012;). The virus initially infects monocytes within lymphoid tissue in the upper respiratory tract and tonsils and is subsequently disseminated via the lymphatics and blood to the entire reticuloendothelial system. Direct viral destruction of a significant proportion of the lymphocyte population and especially CD4+ T cells occurs within the blood, tonsils, thymus, spleen, lymph nodes, bone marrow, mucosa-associated lymphoid tissue, and hepatic Kupffer cells. Massive destruction of lymphocytes results in an initial lymphopenia and transient fever which occurs a few days after infection. Subsequently, there is a second stage of cell associated viremia and fever (8 to 9 days after infection), after which CDV infects cells of the respiratory, gastrointestinal tract, central nervous system (CNS), urinary tract, and skin, as well as red and white blood cells, including additional lymphoid cells (Appel, 1970; Beineke *et al.*, 2009). In this stage of infection, CDV infects a variety of cell lineages including epithelial, mesenchymal, neuroendocrine, and hematopoietic cells, and forms intra-cytoplasmic as well as intra-nuclear inclusions. Infection of epithelial cells by the virus is inefficient and occurs relatively late in the course of infection involving nectin-4 receptor which is also thought to be an exit receptor (Noyce *et al.*, 2013). CDV is shed in all secretions and excretions from the body as early as day 5 after infection before the onset of clinical signs. Shedding of virus can continue for as long as 3 to 4 months, but usually resolves after 1 to 2 weeks.

Clinical Signs

- ☆ *Diphasic fever*: 1st stage-up to 96 hrs then come to normal level till 11-12 days. 2nd stage of viraemia starts from 11-12 days (Systemic infection).
- ☆ *Pulmonary Form*: most prevalent form, coryza like syndrome
- ☆ *Digestive form*: vomiting, abdominal pain, bloody dirrhoea
- ☆ *Ocular form*: Swollen eyelid, conjunctivitis, transient keratitis
- ☆ *Cutaneous Form* (*Hard pad Disease*): rash, vesicle and pustules. Skin of foot pads and nose become hard due to hyper keratitis.

- ☆ *Nervous Form* (*Old Dog Encephalitis*): Seizures and epileptic fits-chewing gum, Cerebellar and vestibular signs, paraparesis or tetraparesis with sensory ataxia and Chorea (Jerky movements of group of muscle). Generally, occurs between two to four weeks after infection.

Diagnosis

The initial diagnosis of CDV is mostly based on identifying the clinical signs that are associated with an infection. However, this form of diagnosis remains problematic and difficult as the clinical presentations of the disease is varied. Differentiation from other diseases with respiratory, neurological and/or gastrointestinal signs, such as rabies, feline panleukopenia, coronavirus, toxoplasmosis, bacterial enteritis and parvovirus, should be conducted. Intoxicants such as lead or organophosphates can also cause simultaneous gastro-intestinal and neurologic signs and must be considered for differential diagnosis.

Sample Collection and Transportation

The ante-mortem diagnosis of canine distemper based on clinical signs is challenging because of the other diseases that mimic the clinic signs. Laboratory testing is required to confirm a suspicion of canine distemper. Submission of the optimal samples assures that the laboratory can produce reliable results for CDV testing. For antigen detection (FAT, RT-PCR, or virus isolation), the optimum time for sample collection is during the height of the febrile response when nasal discharge is still serous or sero-mucoid. This is also the best time to collect the "acute" sample for paired serology. The choice of test will depend on the sample submitted, condition of the sample, sensitivity of the test and the desired turn-around time. Following are the optimal samples for ante-mortem diagnosis:

1. *EDTA-blood*: The distemper virus has a high tropism for white blood cells. Therefore, buffy coat cells are a very good sample for CDV diagnosis. Transport the sample in cold chain but do not freeze it.
2. *Conjuctival or nasal swab*: A conjunctival or nasal swab placed in viral transport medium is an ideal sample for RT-PCR and virus isolation. If viral transport medium is unavailable, a swab moistened with sterile saline (PBS or NSS) and placed in a sealed sterile tube is a good substitute. Transport the sample in cold chain on ice.
3. *Serum*: Detection of CDV antibodies in serum is useful for diagnosis in unvaccinated animals. Even in this group, paired serum samples need to be submitted - one collected at presentation and one collected 2-3 weeks later. A four-fold or greater rise in antibody titer between the "acute" and "convalescent" serum samples confirms the diagnosis of CDV. Serum antibody titers are also useful for ascertaining the vaccination status of animals. Transport the sample in cold chain

4. *Cerebrospinal fluid*: Antibody molecules are normally too large to penetrate the blood-brain barrier so antibodies resulting from vaccination or non-CNS infection will not be detectable in CSF. Positive antibody titers in CSF can confirm the diagnosis of canine distemper. The virus can also be detected in CSF using RT-PCR. Transport the sample in cold chain.
5. *Tissues and other materials*: Various specimens including urinary epithelium cells, skin and stomach biopsy specimens, cells from tracheal washings, blood smears have been used for an etiological diagnosis. The lung, liver, lymph nodes, brain, and spleen of any dead animal with suspected CDV infection are used for viral isolation and/or PCR. The following sets of each tissue sample are required for providing diagnosis:
   - ✰ One set of the tissue samples is chilled but not frozen for immunofluorescence assay.
   - ✰ One set of the tissue samples is collected aseptically and stored frozen at -20°C for virus isolation
   - ✰ One set of tissues is preserved in 10 per cent neutral buffered formalin solution. Such tissue samples are generally used for histopathological examination.

Diagnostic Methods

One of the several techniques that have been developed for the detection of CDV is Reverse Transcriptase-Polymerase chain reaction (RT-PCR). It represents a sensitive and specific method for the early and safe ante-mortem diagnosis of distemper by using serum, whole blood, and/or CSF regardless of clinical signs, pathological findings, neutralizing antibody titers and virus antigen distribution (Frisk *et al.*, 1999; Castilho *et al.*, 2007). Sensitivity of RT-PCR varies depending on the sample source, extraction method and choice of primers (Elia *et al.*, 2006). Various other methods like nested PCR, real-time PCR, hemi-nested multiplex PCR, multiplex PCR and ARMS-PCR are also being used to detect the CDV infection (Kapil *et al.*, 2011; Loots *et al.*, 2017). Commercially available quantitative RT-PCR can usually distinguish natural infection from vaccine virus. A combined two-step RT-PCR to distinguish vaccine strains from emerging wild-type strains has also been described which can be of particular value in epidemiologic investigations or in outbreaks in non-canine species.

Diagnostic tests like immunofluorescence, antigen immunocapture enzyme-linked immunosorbent assay, immunocytochemistry, and *In-situ* hybridization have been used for detection of CDV antigen and CDV RNA. However, the majority of these methods are laborious and time-consuming and more importantly, they are of limited usefulness when they are applied to clinical specimens. Although immunohistochemistry represents a highly sensitive and specific method for detection of CDV antigen in tissue obtained postmortem, it is suitable only within limits for the diagnosis of distemper in living animals (Kapil *et al.*, 2011). Serological assays to detect and determine specific titers against CDV include indirect fluorescent antibody test (IFAT), enzyme-linked immunosorbent assay (ELISA) and the serum-neutralization test. The determination of CDV-neutralizing antibodies in serum or CSF may be helpful in some animals with chronic CNS infection (Soma *et al.*, 2003), but again, the results are variable and depend on the stage of the disease. Serology as a diagnostic test is, however, not reliable in distinguishing between naturally acquired CDV infection (wild-type CDV strain) and infection with attenuated virus vaccine strain or immune response to recombinant, virus-vectored vaccine and should thus be combined with other techniques, such as RT-PCR and viral antigen ELISA. The ELISA and IFAT tests are most commonly used to detect the IgG and IgM levels against CDV (Greene and Appel, 1984).

Ever since the development of Hybridoma Technology in 1975 by Kohler and Milstein, the vision for antibodies as tools for research in animal disease diagnosis have been revolutionized. The monoclonal antibodies are directed against single epitopes are highly specific in nature and in addition these can be produced in huge quantities without the use of animals. Monoclonal antibodies have been deployed in different tests like ELISA and lateral flow assays for detection of viral antigen in tissues, secretions (conjuctival and nasal swabs), lymphocytes and blood by various workers in different tests (Potgeiter*et al.*, 1989; An *et al.*, 2008; Li *et al.*, 2012; Zhang *et al.*, 2020). Another approach of using mAb based ELISA for antigen detection is *cell-ELISA* which is a very simple, sensitive, rapid, inexpensive assay that has been used for detection and quantification of virus infected cells using monoclonal antibody directed against one of the viral proteins. The cell-ELISA has been used for quantification of molecules that are being expressed on the surface of cells (Ogino *et al.*, 2003).

CDV has been isolated in various cell culture systems. A virulent CDV can be easily cultured in macrophages or lymphocytes; however, it grows only with adaptation in epithelial or fibroblast cell lines. Macrophage cultures have been replaced by dog lymphocyte cultures for the isolation of CDV. Buffy coat cells or tissues from infected animals can be cultivated with mitogen stimulated canine blood lymphocytes, and cultures are examined 72 to 144 hours later by immunofluorescence (Appel*et al.*, 1992). A marmoset lymphoid cell line (B95a) has also been used for CDV isolation (Kai *et al.*, 1993). Canine distemper virus has been isolated effectively using cell lines such as MDCK (canine epithelial kidney cells), MV 1 LU (mink lung), and Vero (African green monkey kidney cells) (Lednicky*et al.*, 2004a). Vero cells expressing canine signaling lymphocyte activation molecule (Vero-dogSLAM) has been established to aid the isolation of CDV from clinical material within 24 hours of inoculation (Seki *et al.*, 2003). Of late, Vero-dog Nectin cell line has also been used for isolating the virus (Noyce *et al.*, 2013). However, the virus isolation being time consuming procedure is not suitable for diagnosis of clinical samples.

Prophylaxis

Currently the best means for breaking the circulation of CDV between susceptible wildlife populations and domestic dogs is through regular vaccination of pet dogs and preventing them from roaming freely and interacting with unvaccinated dogs and wildlife that may harbor the virus. Modified live vaccines (MLV) against CDV were introduced in early 1960s. Current licensed CDV vaccines contain viral strains that have been attenuated by serial passage, either on canine kidney cells (Rockborn), hen eggs (original Onderstepoort) or chicken fibroblast cultures (Lederle strain). These vaccines have significantly reduced CDV infection in domestic dog population, although MLV induced outbreaks have been reported in the wildlife population.

A recombinant CD vaccine, incorporating the F and H CDV proteins in the canarypox virus strain was developed. This vaccine was found to be safe to all prone species tested to date, together with European ferrets (*Mustela putoriusfuro*), dogs, fennec foxes, giant pandas (*Ailuropoda melanoleuca*), meerkats (*Suricata suricatta*) and Siberian polecats (*Mustela eversmanni*) (Loots *et al.*, 2017). Whereas the canarypox vectored vaccine is safe in abundant target species, it is replication-defective and as such induces a milder immunological response than MLV vaccines (Schultz *et al.*, 2006). However, this final trait highlights its applicability for immunization of juvenile animals, when maternal antibodies are present (Pardo *et al.*, 2007).

Oral bait vaccines to control wildlife distemper are not yet available. Two major issues in developing an efficacious oral bait vaccine for distemper are: achieving an adequate mucosal immune response in the gut and overcoming interference from maternal antibodies in infant animals. Attempts at inducing mucosal immunity using vaccinia and canarypox vectored CDV vaccines have been reported using ferrets as model animals.

Conclusion

Availability of diagnostic assays for precise disease diagnosis, infrastructure required for mass scale testing and vaccines are the elementary to any disease control programme. CD has battery of diagnostics blend with conventional to modern test. Conventional virology techniques like virus isolation and virus neutralization test are still in use for laboratory diagnosis because of their high specificity. Molecular assays like RT-PCR, real time RT-PCR and LAMP are used for their rapid diagnosis. Immunoassays like ELISA based on IgM and IgG antibodies are commonly used to detect recent infection and in later phase of disease, respectively. Several pen-side tests or point of care test (POCT) have been in-use for the diagnosis of ailing animals and accessing the antibodies. These tests give rapid results within 10 min and works efficiently without involvement of sophisticated instruments and skilled person. Use of effective vaccine is another key component for any disease to control. MLV is such a conventional vaccine being in place for containing the CDV infection, especially in dogs. Use of engineered vaccines like vectored vaccines is also being used by several countries with more use in wildlife population. In spite of vaccination, outbreaks are seen in vaccinated population which is a worth consideration and need further investigation.

Acknowledgement

The authors are thankful to the Science and Engineering Research Board-Department of Science and Technology (SERB-DST) Government of India grant No. EEQ/2018/000823 for financial assistance.

References

1. An, D. J., Kim, T. Y., Song, D. S., Kang, B. K. and Park, B. K. 2008. An immunochromatography assay for rapid antemortem diagnosis of dogs suspected to have canine distemper. J Virol Methods. 147: 244–249.
2. Appel, M. J. 1970. Distemper pathogenesis in dogs. Journal of the American Veterinary Medical Association. **156**: 1681-1684.
3. Appel, M. J. G. and J. H. Gillespie. 1972. Canine distemper virus. *In*: Gard, S., Hallauer, C. and Meyer, K. F. ed. Virology monographs 11. Springer-Verlag, New York. pp. 1–96.
4. Appel, M. J. G., Pearce-Kelling, S. and Summers, B. A. 1992. Dog lymphocyte cultures facilitate the isolation and growth of virulent canine distemper virus. Journal of Veterinary Diagnostic Investigation. **4:** 258-263.
5. Beineke, A., Baumgartner, W., Wohlsein, P. 2015.Cross-species transmission of canine distemper virus – an update.One Health.**1**: 49–59.
6. Beineke, A., Puff, C., Seehusen, F., Baumgartner, W. 2009.Pathogenesis and immunopathology of systemic and nervous canine distemper. Vet Immunol Immunopathol. **127**: 1–18.
7. Budaszewski, R. F., and von Messling, V. 2016. Morbillivirus Experimental Animal Models: Measles Virus Pathogenesis Insights from Canine Distemper Virus. Viruses, **8**(10), 274.
8. Castilho, J. G., Brand~ao, P. E., Carnieli, P., Oliveira, R. N., Oliveira, C. I., Macedo, Z. M. P., Peixoto, M. L., Carrieri, I., Kotait. 2007. Molecular analysis of the N gene of canine distemper virus in dogs in Brazil. Arq Bras Med Vet e Zootec. **59**: 654–659.
9. Cleaveland, S., Appel, M. G., Chalmers, W. S., Chillingworth, C., Kaare, M., Dye, C. 2000. Serological and demographic evidence for domestic dogs as a source of canine distemper virus infection for Serengeti wildlife. Vet Microbiol. **72**: 217–227.
10. Deem, S. L., Spelman, L. H., Yates, R. A. and Montali, R. J. 2000. Canine distemper in terrestrial carnivores: a review. J. Zoo and Wildlife Med. 31(4): 441–451.

11. Elia, G., Decaro, N., Martella, V., Cirone, F., Lucente, M. S., Lorusso, E., Di, Trani, L., Buonavoglia, C. 2006. Detection of canine distemper virus in dogs by real-time RT-PCR. J Virol Methods. **136**: 171–176.

12. Frisk, A. L., König, M., Moritz, A. and Baumgärtner, W. 1999. Detection of canine distemper virus nucleoprotein RNA by reverse transcription-PCR using serum, whole blood, and cerebrospinal fluid from dogs with distemper. Journal of Clinical Microbiology.**37**: 3634-3643.

13. Greene, C. E. and Appel, M. J. 1984. Canine distemper. *In*: Clinical microbiology and infectious disease of the dog and cat. Edited by Greene, C. E. Saunders, Elsevier. pp. 386-405.

14. Greene, C. E. and Appel, M. J. 2006. Canine Distemper. *In*: GREENE Infectious diseases of the dog and cat (3rd edn). Edited by Greene, C. E: Saunders, Elsevier.pp. 25-41.

15. Kai, C., Ochikubo, F., Okita, M., Linuma, T., Mikami, T., Kobnne, F. and Yamanouchi, K. 1993. Use of B95a cells for isolation of canine distemper virus from clinical cases. Journal of Veterinary Medical Sciences. **55**: 1067-1070.

16. Kapil, S., Yeary, T. J. 2011. Canine distemper spillover in domestic dogs from urban wildlife. Vet Clin Small Anim. **41**: 1069–1086.

17. Kennedy, S., Kuiken, T., Jepson, P. D., Deaville, R., Forsyth, M., Barrett, T., van de Bildt, M. W., Osterhaus, A. D., Eybatov, T., Duck, C., Kydyrmanov, A., Mitrofanov, I. and Wilson, S. 2000. Mass die-Off of Caspian seals caused by canine distemper virus. Emerg Infect Dis. **6**: 637–639.

18. Lamb, R. A. and Kolakofsky, D. 2001. Paramyxoviridae: The viruses and their replication. *In*: Fundamental Virology, 4th edn. Edited by Knipe, D. M. and Howley, P. M. Philadelphia: Lippincoot Williams and Wilkins. pp. 689-724.

19. Lednicky, J. A., Meehan, T. P., Kinsel, M. J., Dubach, J., Bocchetta, M., Hungerford, L. L., Sarich, N. A., Witeki, K. E., Braid, M. D., Pedrak, C. and Houde, C. M. 2004a. Effective primary isolation of wild-type canine distemper virus in MDCK, MV1 Lu and Vero cells without nucleotide sequence changes within the entire haemagglutinin protein gene and in subgenomic sections of the fusion and phosphoprotein genes. Journal of Virological Methods. **118**: 147-157.

20. Liu, Y., Hao L., Li X., Wang, L., Zhang, J., Deng, J., and Tian, K. 2017. Development and characterization of canine distemper virus monoclonal antibodies. Monoclon Antib Immunodiagn Immunother. 36: 119–123.

21. Loots, A. K., Mitchell, E., Dalton, D. E., Kotzé, A., Venter, E. H. 2017. Advances in canine distemper virus (CDV) pathogenesis research: a wildlife perspective. Journal of General Virology.**98**: 311–321.

22. Mamaev, L. V., Denikina, N. N., Belikov, S. I., Volchkov, V. E., Visser, I. K., Fleming, M., Kai, C., Harder, T. C., Liess, B., Osterhaus, A. D., Barrette, T. 1995. Characterisation of morbilliviruses isolated from Lake Baikal seals (Phocasibirica). Vet Microbiol. **44**: 251–259.

23. Mühlebach, M. D., Mateo, M., Sinn, P. L., Prüfer, S., Uhlig, K. M. *et al.*, Adherens junction protein nectin-4 is the epithelial receptor for measles virus. Nature. **480**: 530–533.

24. Noyce, R. S., Delpeut, S., Richardson, C. D. 2013. Dog nectin-4 is an epithelial cell receptor for canine distemper virus that facilitates virus entry and syncytia formation. Virology. **436**: 210–20.

25. Ogino, T., Wang, X., Ferrone, S.2003. Modified flow cytometry and cell-ELISA methodology to detect HLA class I antigen processing machinery components in cytoplasm and endoplasmic reticulum. J Immunol Methods.278: 33–44.

26. Pope, J. P., Miller, D. L., Riley, M. C., Anis, E., Wilkes, R. P. 2016. Characterization of a novel Canine distemper virus causing disease in wildlife. J Vet Diagn Invest. **28**(5): 506-13.

27. Potgieter, L. N. and Ajidagba, P. A. 1989. Quantitation of canine distemper virus and antibodies by enzyme-linked immunosorbent assays using protein A and monoclonal antibody capture. J Vet Diagn Invest. 1: 110–115.

28. Pratakpiriya, W., Seki, F., Otsuki, N., Sakai, K., Fukuhara, H., Katamoto, H., Hirai, T., Maenaka, K., Techangamsuwan, S., Lan, N. T., Takeda, M., Yamaguchi, R. 2012.

29. Ramanathan, A., Malik, P. K., Prasad, G. 2007. Seroepizootiological survey for selected viral infections in captive asiatic lions (*pantheraleopersica*) from western India. Journal of Zoo and Wildlife Medicine.**38**(3): 400-408.

30. Schultz, R. D. 2006. Duration of immunity for canine and feline vaccines: a review. Vet Microbiol. **117**(1): 75–9.

31. Seki, F., Ono, N., Yamaguchi, R., and Yanagi, Y. 2003. Efficient Isolation of Wild Strains of Canine Distemper Virus in Vero Cells Expressing Canine SLAM (CD150) and Their Adaptability to Marmoset B95a Cells. Journal of Virology. **77**(18): 9943–9950.

32. Soma, T., Ishii, H., Hara, M., Ohe, K., Hagimori, I., Ishikawa, Y. and Taneno, A. 2003. Detection of canine distemper virus antigen in canine serum and its application to diagnosis. Veterinary Record**.** **153**: 499-501.

33. Tatsuo, H., Ono, N., Yanagi, Y. 2001. Morbilliviruses use signaling lymphocyte activation molecules (CD150) as cellular receptors. J Virol. **75**: 5842–5850.

34. Viana, M., Shirima, G. M., John, K. S., Fitzpatrick, J., Kazwala, R. R., Buza, J. J., Cleaveland, S., Haydon, D. T., and Halliday, J. 2016. Integrating serological and genetic data to quantify cross-species transmission brucellosis as a case study. Parasitology, 143(7), 821–834.

35. Zhang, Y., Xu, G., Zhang, L., Zhao, J., Ji, P., Li, Y., Liu. B., Zhang, J., Zhao, Q., Sun, Y. and Zhou, E. 2020. Development of a double monoclonal antibody–based sandwich enzyme-linked immunosorbent assay for detecting canine distemper virus. Applied Microbiology and Biotechnology. 104: 10725–10735.

Chapter 16

Recent Trends in Diagnosis of *Parvo Virus* Infection

Shubhangi Warke

I/c Professor and Head, Department of Veterinary Microbiology and Animal Biotechnology T and R Cell, Nagpur Veterinary College, Nagpur, Maharashtra

Canine parvovirus -2 is a highly infectious viral disease of dogs and one of the serious concerns to pet owners, practicing veterinarians and scientists due to its high morbidity and mortality rate. CPV-2 is stable, highly contagious virus which mainly affecting canine population.CPV-2 spread by direct or indirect contact with infected faeces of dogs. Vaccine against *Canine parvovirus* prevents this infection but case fatality can reach up to 91 per cent if cases are untreated.

During late 1970s, a new contagious viral disease of pups emerged showing gastroenteritis or myocarditis like symptoms throughout the world, named as CPV-2. It was first recognized in 1977 and described in several countries in mid 1978, then it has been established as a whole pathogen of dogs throughout the world. After a time of adaptation, the virus begin to be highly contagious for dogs, resulting in the prevalent and apparent in 1978–1979. CPV-2 virus has 98 per cent similarity with *feline panleukopenia*, only differ in two amino acids present in the viral capsids proteins VP2 and also homogeneous to Mink enteritis and parvovirus of Foxes and Raccoons.

Canine parvovirus belongs to family *Parvoviridae* comprising two subfamilies *Parvovirinae* and *Densovirinae*. *Parvovirinae* contain virus of vertebrates and *Densovirinae* contain virus of insects. There are 3 genera in sub family *Parvovirinae*viz, *Parvovirus Erythrovirus* and *Dependovirus*. (Murphy *et al.*, 1999). Sequence analysis has also divulged that CPV strains have undergone a series of evolutionary selections in nature, which was globally distributed as new variants which have completely replaced the original CPV-2. Recently, CPV has three main antigenic variants *viz.* 2a, 2b and 2c which are distributed among the dog population throughout the world. (Decaro*et al.*,2006). Previously, CPV-2c was nominated as CPVGlu-426 mutant that originated initially from Italy, but now has been detected in other countries also (Buonavoglia*et al.*, 2001). The CPV-2 has spread in identical pattern and the disclosure and subsequent renewal by the new antigenic strains has also been shown by antigenic analysis of isolates from Germany, France, Denmark, Japan, Italy, Australia, Spain and Africa although the proportions of CPV-2a, CPV-2b and CPV-2c strains has been differ. (Parrish *et al.*,1988, Mochizuki *et al.*,1993, Ybanez*et al.*,1995, Truyen*et al.*,1996, Steinel*et al.*,1998, Buonavoglia*et al.*, 2001). These viruses have become distributed globally and there antigenic as well as genetic variation in the virus strains have also been corresponding with changes occurs in host range of the virus (Parrish *et al.*, 1999). Phylogenetic analysis has divulged that CPV strains have undergone a sequence of progressive selections in nature, which have resulted in worldwide distribution of current variants have been replaced the original CPV-2.

Canine parvovirus-2 are non enveloped having 25nm in diameter with icosahedral symmetry. Genome consist of single stranded negative sense DNA virus having size of 5.2 kb in length (Murphy *et al.*,1999). Terminal palindromic sequences at each end to form hairpin structure.CPV-2 genome has 2 open reading frame (ORF) which are further divided into ORF 1 and ORF 2.The ORF1 encodes for two nonstructural protein NS1 and NS2 that are translated through alternate splicing of the viral mRNAs and ORF 2 encodes for two capsid protein *i.e.* VP1 and VP2.

The disease is widespread prevalent in Asia (Khadilkar*et al.*, 1994). In India, prevalence of canine parvovirus was first announced by Balu and Thangaraj in 1981 in Madras and Ramadas and Khader in 1982, confirmed that CPV-2 is an etiological agent of *Canine parvovirus* infection. After that enormous number of outbreaks of *Canine parvovirus* have been noted in various clinic from various empire of India like Kerala, Tamil Nadu, Karnataka, Andhra Pradesh, Odisha,Pondicherry (Mukhopadhyay *et al.*,2014), Haryana (Savi*et al.*,2009) and Uttar Pradesh (Nandi *et al.*,2010, 2010a).

The virus is transmitted via oro-fecal route and requires rapidly dividing cells of fetuses and new born or of hematopoietic and intestinal tissues of young and adults animals. (Decaro and Buonavoglia., 2012).

Although severe clinical disease typically occurs in dogs younger than 6 months of age, adults with insufficient immunity may potentially be affected. (Kalli *et al.*, 2010 and Marcovich*et al.*, 2012). Breed predisposition and seasonal prevalence of the disease are subject to considerable geographic variation (Kalli *et al.*, 2010). CPV-2 is ubiquitous and can survive in the environment for more than a year, enabling exposure of susceptible dogs to infected feces, vomitus, or fomites(Sykes,2014). The incubation period following natural or experimental exposure ranges from 4 to14 days, and virus shedding starts a few days prior to the occurrence of clinical signs, progressively declining 3–4 weeks post exposure. (McCaw and Hoskins.,2006)

VP2 which is a major capsid protein of CPV, play a major role in the diagnosis of host range and antigenicity of CPV. Mutation which affect VP2 gene are predominantly responsible for evolving different antigenic variant of CPV-2.(Phromnoi*et al.*,2010,Kaur *et al.*,2015).The constant emergence of new strain of CPV-2 is a big concern of dog owners, veterinarian around the world.

Combined vaccine of CPV contains an inactivated and modified virus. These vaccine help to prevent CPV, but when there is emergence of newer antigenic variant, vaccine get fail to protect pups against CPV, so identification along with the knowledge of genetic variation of VP2 can be help in detecting the new strain of CPV and it help in development of new vaccine.

Diagnosis of disease have been done by using various methods like Haemagglutination (HA) Haemagglutination inhibition test (HI),Rapid antigen detection kit, agar gel precipitation test (AGPT),Enzyme linked immunosorbent assay (ELISA),electron microscopy, indirect florescent antibody test, virus isolation by cell culture using Madindarby canine kidney (MDCK),Crandle feline kidney (CRFK) and A-72 cell lines. Among them, virus isolation done by using various cell lines is highly specific way to identification, HA and HI are simple, conventional method of diagnosis. For electron microscopy, concentrated virus to be required. ELISA using monoclonal antibodies is also the another way for diagnosis of CPV. Restriction Enzyme (RE) digestion, real time PCR, Loop-mediated isothermal amplification (LAMP), Nucleic acid hybridization or dot blot, in situ hybridization, nucleic acid sequencing.Among all conventional method, molecular assay like PCR,nested PCR are highly specific and sensitive way of diagnosis. Real time PCR is also sensitive and specific technique for parvoviral DNA.

Conclusion

Currently, it has been observed that the *Canine parvovirus* serotypes circulating in the Indian landmass is completely different from that included in the vaccine and may leading to decrease in the efficacy of the CPV-2 build vaccines. (Greenwood *et al.*, 1995., Yule *et al.*, 1997, Pratelli*et al.*, 2001) could be the one reason for the infection in the vaccinated pups (Savi*et al.*, 2009). Vaccine which are used in India against *Canine parvovirus* is inactivated, modified live vaccine which protect the pups, but recent study observed that the *Canine parvovirus* circulating in India is entirely different from strain included in the vaccine, so there is failure of vaccination observed, because some of the study noticed that, proportion of parvoviral infection in vaccinated and un vaccinated dog is nearly same, these may be the reason for the infection in the vaccinated pups.(Savi*et al.*, 2009).Considering this problem, it was observed strain used in vaccine that were isolated about 25 to 30 year ago (Nandi *et al.*, 2010). On the other hand, further research may be warranted in elucidating to which extend the apparent vaccination failures in the clinical setting are vaccine-associated (*e.g.*, vaccines with reduced immunogenicity against the new field variants) or vaccination policy-associated (*e.g.*, level of herd immunity in an area, schedule of primary vaccination series, booster timing).

Most of the stray dogs are unvaccinated against the disease and they become the origin of disease to others susceptible host, early diagnosis of CPV infection by using conventional as well as molecular methods help to protect dogs.

References

1. Balu, P. A and T.M.Thangaraj (1981) Canine viral gastroenteritis- A clinical report. Ind.J.Vety. Med. 1: 73.
2. Buonavoglia, C., V.Martella, A.Pratelli, M. Tempesta, A.Cavalli, D. Buonavoglia, G.Bozzo, G.Ella, N. Decaro and L. Carmichael (2001) Evidence for evolution of canine parvovirus type 2 in Italy. J. Gen. Virol. 82(12): 3021-25.
3. Decaro, N.and C.Buonavoglia (2012) Canine parvovirus-A review of epidemiological and diagnostic aspects with emphasis on type 2c.Vety. Microbiol.155: 1-12.
4. Decaro, N., G. Elia, V.Martella, M.Campolo, C.Desario, M.Camero (2006) Characterisation of the *Canine parvovirus* type 2 variants using minor groove binder probe technology. J. Virol. Methods.133: 92–9.
5. Greenwood, N, M., Chalmers, W, S, K., Baxendale, W and Thompson, H (1995) Comparison of isolates of *Canine ParvoVirus* by restriction enzyme analysis and vaccine efficacy against field strains. Vet Rec, 136: 63-67.
6. Kalli I, Leontides LS, Mylonakis ME, Adamama-Moraitou K, RallisT, Koutinas AF.(2010) Factors affecting the occurrence, duration of hospitalization and final outcome in *Canine parvovirus* infection. *Res VetSci.*; 89(2): 174–178.
7. Kalli I, Leontides LS, Mylonakis ME, Adamama-Moraitou K, RallisT, Koutinas AF. (2010) Factors affecting the occurrence, duration of hospitalization and final outcome in canine parvovirus infection. *Res VetSci.*; 89(2): 174–178.
8. Kaur, G., M.Chandra, P. N. Dwivedi and N.S. Sharma (2015) Prevalence of *Canine parvovirus* in dogs in Ludhiana, Punjab. Res. Environ. Life Sci. 8(2): 157-158.
9. Khadilkar, M. V., P.Batra and N.M.Singhvi (1994) *Canine parvovirus*. J. Remount Vety. Corps.33: 81.
10. Marcovich JE, Stucker KM, Carr AH, Harbison CE, Scarlett JM, Parrish CR.(2012) Effects of *Canine parvovirus* strain variations on diagnostic test results and clinical management of enteritis in dogs. *J Am Vet MedAssoc*.241(1): 66–72.

11. Markovich, J., Stucker, K., Carr, A., Harbison, C., Scarlett, J., Parrish, C. (2012) Effects of *Canine ParvoVirus* strain variations on a diagnostic test of enteritis in a dog. J. Amer. Vety. Med. Asso.241: 66–72.
12. McCaw DL and Hoskins JD.(2006) Canine viral enteritis. In: Green CE, editor.*Infectious Diseases of the Dog and Cat*. 4th ed. St Louis, MO: Saunders; 63–73.
13. Balu, P. A and T.M.Thangaraj (1981) Canine viral gastroenteritis- A clinical report. Ind.J.Vety. Med. 1: 73.
14. Mochizuki, M., M.C.San, Gabriel, H.Nakatani, M.Yoshida, and R. Harasawa(1993) Comparison of polymerase chain reaction with virus isolation and haemagglutination assays for the detection of *Canine parvoviruses* in faecal specimens. Research in Vety. Sci. 55: 60-63.
15. Mukhopadhyay, H.K., S.Lakshmi Matta, S. Amsaveni, P.X. Antony, J.Thanislass and R.M. Pillai (2014) Phylogenetic analysis of *Canine parvovirus* partial VP2 gene in India. Virus Genes. 48: 89-95.
16. Murphy, F.A., E.P.J.Gibbs, M. Horzinek and M.J. Studdert (1999) Veterinary Virology. 3rd ed. New York: Academic Press. 343-354.
17. Nandi, S., R. Anbuzhagan and Manoj Kumar (2010) Strain differentiation and characterization of *Canine parvovirus* by PCR and RE mapping. Ind.J.Biotechnol. (9): 38-42.
18. Parrish, C. R. (1999) Host range relationship and the evolution of *Canine parvovirus.*Vety.Microbiol. 69: 29–40.
19. Parrish, C.R., G.Burtonboy, L.E.Carmichae (1988) Characterization of a nonhemagglutination mutant of *Canine parvovirus*. Virology. 163: 230-232.
20. Phromnoi, S., R.Sinsiri and T. Sirinarumitr (2010) Expression of Recombinan VP2 Protein of *Canine Parvovirus* in *Escherichia coli.*Kasetsart Journal (Natural Science) 44: 870 -78.
21. Pratelli, A., A.Cavalli, V.Martella, M.Tempesta, N.Decaro, L.E.Carmichael and C.Bounavoglia (2001) *Canine parvovirus* (CPV) vaccination: comparison of neutralizing antibody responses in pups after inoculation with CPV-2 or CPV-2b modified live virus vaccine. Clin.Diagn. Lab. Immunol. 8: 612-15.
22. Ramadass, P and T.G.A.Khader (1982) Diagnosis of *Canine parvovirus* infection by agar gel precipitation test and fluorescent antibody techniques. Cheiron 11: 323-25.
23. Savi, J., P. Minakshi and G. Prasad (2009) Genotyping of field strains of canine parvovirus in Haryana using PCR and RFLP. Ind. J. Ani. Sci.79(10): 971-97.
24. Steinel, A., E.H.Venter, M.Van Vuuren, C.R.Parrish and U.Truyen (1998) Antigenic and genetic analysis of *Canine parvoviruses* in southern Africa. OnderstepoortJ.Vety.Res. 65(4): 239-42.

25. Sykes JE.(2014) *Canine parvovirus* infections and other viral enteritides.In: Sykes JE, editor. *Canine and Feline Infectious Diseases*. 1st ed.St Louis, MO: Elsevier; : 141–151.

26. Truyen, U., J.F.Evermann, E.Vieler and C.R.Parrish (1996) Evolution of *Canine parvovirus* involved loss and gain of feline host range. Virology. 215: 186-89.

27. Yule, T. D., M.B.Roth, K.Dreier, A.F. Johnson, M.Palmar-Densmore, K.Simmons and R.Fanton (1997) *Canine parvovirus* vaccine elicits protection from the inflammatory and clinical consequences of the disease. Vaccine. 15: 720-22.

Chapter 17

Recent Update of Footrot in India

S. Vijayalakshmi

Associate Professor,
Department of Microbiology CVSc, Tirupati, Andhra Pradesh

Footrot is a contagious hoof disease of sheep and other ungulates and begins as an interdigital dermatitis, which is followed by formation of lesions on the interdigital wall of the hoof and subsequent separation of the hard horn from the foot with strong odour. The essential transmissible agent of the disease is the bacterium *Dichelobacternodosus* (*D.nodosus*), although the role of other infective agents in the onset of disease is not fully understood. However recent studies suggest *D. nodosus* being the primary pathogen (Witcomb *et al.*, 2014) in establishing the disease followed by invasion of the *Fusobacterium necrophorum* (F. necrophorum).

D. nodosus is a member of *Cardiobacteriaceae* family and grouped with genera *Cardiobacterium* and *Suttonella*. This bacterium is slow growing, fastidious and obligatory anaerobic bacteria. *D.nodosus* is gram negative, rod shaped bacterium with characteristic knobs at each end and is often heavily fimbriated. *D. nodosus* cells are straight or curved rods, 3µm to 6µm X 1.0µm to 1.7µm in size, nonspore forming, non capsulated rods. The current Australian system of classification is based on antigenic variation in structure of fimbriae and divided *D. nodosus* into ten serogroups (A-I and M) and 18 serotypes using slide agglutination and tube agglutination test respectively (Ghimere *et al.*, 1998)

Epidemiology

Sheep and goats are the primary species affected and they are susceptible at all ages. Sheep and goats may develop a severe form of the disease whereas cattle and deer develop a mild form. Footrot has been reported in wild ungulates, ibex (*Capra ibex ibex*) and mouflon (*Ovis aries musimon*) in Germany and Switzerland. Environmental factors affect footrot expression and transmission. Footrot may not occur year round, it shows peaks of clinical expression in moist rainy conditions when there is adequate moisture and ambient temperatures prevail. Suboptimal environmental conditions are not conducive to full clinical expression of virulent footrot and lesions examined under these conditions may be confined to the interdigital skin and present as mild lesion The prevalence of lesions and persistence of infection varies; dry warmer climates tended to result in lower prevalence and decreased persistence, whereas wetter and cooler climates resulted in a higher prevalence of lesions and increased persistence.

In IndiaDarzi *et al.* (2002) first to report the gross and microscopic lesions of footrot in of Kashmir valley. Later Wani *et al.* (2004) confirmed the presence of foot rot in the valley by detecting, *D. nodosus*by PCR using species-specific 16S rRNA primers. Wani et al(2007)isolated and characterized *D. nodosus*from ovine and caprine footrot. Serogrouping of the isolates was accomplished by Hussain *et al.* (2009) with multiplex-PCR, using nine (A-I) serogroup specific primers and reported the presences of 'B', 'I" and 'E" serogroups.

Sreenivasulu*et al.* (2013) first time reported theoccurrence of footrot in sheep in tropical climate of Andhra Pradesh and Tamil Nadu, where temperature reaches to 42°C and later reported overall prevalence of 11.41 per cent in the region. These out breaks reported mostly during North east monsoon period (October to December). Out of 1,050 samples collected from footrot lesions of naturally infected sheep, 478 (45.22 per cent) samples revealed the presence of *D.nodosus*. Results of serogrouping revealed circulation of six different serogroups (A,B,C, E, F and I). Out of 478 positive samples, majority (44.14 per cent) of them carried serogroup B which was followed by serogroup I in 25.94 per cent, A in 12.13 per cent, F in 6.69 per cent and C in 4.60 per cent while serogroup E was detected in only one sample. Thirty (6.27 per cent) samples had mixed infection of two serogroups (A and B or B and C or B and I). Serogroups A and F were detected for the first time in India. Serogroup B was detected in all the districts. Serogroup A was detected in 6, C in 4, I in 5, F in 3 districts, whereas E was detected in only one district of Andhra Pradesh. Serogroups A and I confirmed to the two districts (Chittoor and Nellore) while serogroup F confined to one district (Krishna) only. Virulent isolates of *D.nodosus* across different serogroups were also detected based on the production of thermostable proteases and presence of int A gene.

Vijjayalakshmi *et al.*(2020) characterized the *D.nodosus* isolates by phenotypic virulence test in which Gelatin gel test revealed 56.5 per cent of isolates as virulent, 31.25 per cent of isolates as Intermediate and 12.25 per cent of isolates as benign. Whereas elastase test identified only 18.75 per cent of isolates as virulent, 12.5 per cent of isolates as intermediate, 12.5 per cent of isolates as benign, the rest 56.25 per cent of the isolates did not exhibit any elastase activity. Vinod kumar *et al.* (2016) identified and characterized the 'H' serogroup for the first time in India

The genetic relationships between bacterial strains of the *D.nodosus* was assesed by Multiple-locus VNTR analysis (MLVA). Presence of MLVA types of AP, Telangana and Tamil Nadu isolates at the terminal nodes and J and K and Australian isolates in the internal nodes of minimum spanning tree suggest that footrot was present in Australia and J and K prior to its presence in AP, Telangana and Tamil Nadu. Further it also suggest that occurrence of footrot in this region is of recent one when compared to other countries.

Clinical Description

The clinical manifestation of footrot in sheep is determined by virulence of the strain of *D. nodosus* involved, environmental conditions and the innate and acquired resistance of the host. Interactions between these factors determine the severity of the disease, which is expressed as a continuous spectrum of clinical entities ranging from benign to highly virulent footrot. Three clinical forms of footrot have been described they are: virulent, intermediate and benign. These terms are also being used to describe the strains of *D. nodosus* which are isolated from and have the potential of causing respective clinical forms in exposed, susceptible animals.

Virulent Footrot (VFR)

Virulent footrot in sheep is characterised by initial inflammation of the interdigital skin which is followed by a break in the skin-horn junction on the axial aspect of one or both digits. This inflammatory disruption progresses into the epidermal matrix of the hoof. This results in extensive underrunning, extending up to the anterior and abaxial wall of the hoof horn. On feet which have been affected for several weeks or months, the hoof becomes long and misshapen and lameness is severe. Lesions of VFR have a distinctive foul smelling odour. Clinically footrot is considered virulent if more than 10 per cent of affected animals have under running of hard horn in at least one foot. VFR causes a very significant economic loss to the sheep industry in India

Benign Footrot (BFR)

Benign footrot is often erroneously called footscald. The main lesion of BFR is interdigital dermatitis with little or no under running of the hoof. It causes only mild lameness and rapidly regresses after topical treatment or after the advent of dry environmental conditions. In BFR a large proportion of the flock can be affected but only under conditions favorable for spread of the disease. Clinically BFR is diagnosed if less than 1 per cent of the affected sheep have underrunning of the hard horn of hoof.

Intermediate Footrot (IFR)

The intermediate form of footrot lies between virulent and benign forms with both lesion severity and the production losses observed. Severe lesions may be seen in a few sheep. A variable percentage of the flock may be affected depending on the climatic conditions. Only a small percentage, less than 10 per cent of affected sheep have underrunning lesions. A majority of the animals in the flock have lesions either restricted to the interdigital skin or underrunning of only the soft horn of heel and sole.

Diagnosis of Footrot

Footrot is usually detected by clinical signs, followed by laboratory confirmation of the presence of *D.nodosus* in lesion material. Footrot need to be differentiated from other foot diseases like Traumatic injury, Necrobacillosis, dermatophilosis, Bluetongue, parasitic dermatitis, arthritis and FMD.

1. Clinical Signs

The disease is characterized by an exudative inflammation with a strongly characteristic foul odour, followed by the necrosis of the epidermal tissues of the interdigital skin and hoof matrix, resulting in separation of the hoof from the underlying soft tissues.

2. Laboratory Diagnosis

a) Isolation and Identification of Bacteria

Isolation

D.nodosus can be isolated from interdigital skin/foot lesions of sheep, goats, cattle, deer, or other animals suspected of footrot on solid media. The isolation agar medium inhibits the motility of flagellated bacteria allowing *D.nodosus* to move by twitching away from contaminants and thereby establish a pure colony of *D.nodosus* is most successful from lesions of score greater than 1.

Media and Reagents

For isolation of *D.nodosus*, the commonly used media is the TAS (Trypticase arginine serine medium) supplemented with 2-4 per cent hoof powder. The TAS medium consists of Tryptose (1.5gm), Peptone (0.5gm), Beef extract(0.5gm), yeast extract (0.2gm). Arginine (0.5gm), $MgSO_4$ (0.2gm), serine (0.15gm), hoof powder (2-4 per cent) and H_2O (100ml).

Inoculation

For isolation of *D. nodosus*, the swab samples collected in field from clinical cases should preferably be streaked at the place of collection. However, if the inoculation is not feasible at the place of collection, the samples may be transported in anaerobic transport medium and should be carried to laboratory on ice.

Identification

The Preliminary identification of *Dichelobcaternodosus* is based on colony morphology and grams reaction. On gram reaction the *D.nodosus* appears as Gram negative rods with one or both ends bulged. Pathogenic colonies appear as beaded or papillated with a centre followed by middle zone and peripheral zone. Less pathogenic colonies are mucoid.

Molecular Characterization of *D.nodosus* is carried out by detection of nucleic acid genome by PCR using 16 s r RNA gene and Serogrouping by multiplex PCR using serogroup specific primers (A-I and M) by targeting *fim A* gene.

Control

Control of footrot is mainly by following maintaining good hygienic conditions in herd and minimizing predisposing factors. Foot baths containing 5 per cent copper sulphate or 10 per cent Zinc sulphate or 5- 10 per cent formalin will be of some use. Though vaccination is one of the best methods to control the footrot, vaccines are not available commercially for use in the Country.

Inactivated Trivalent Vaccine

The trivalent inactivated vaccine using the indigenous strains 'B', 'I' and 'A' serogroups was developed at State Level Diagnostic Laboratory, SVVU, Tirupati by Raniprameela *et al.* (2016). Laboratory and field trials were successfully completed and the vaccine found to elicit satisfactory protective immune responses in animals. Now vaccine is ready for release and commercialization. Development of recombinant protease and recombinant fimbrial vaccines is under progress in Department of Microbiology, C.V.Sc., Tirupati.

References

1. Cheetham B F, Tanjung L R and Sutherland M 2006 Improved diagnosis of virulent ovine footrot using the *intA* gene. *Vet Microbiol;* 116(1-3): 166–174.
2. Ghimire S C, Egerton J R, Dhungyel O P and Joshi H D 1998 Identification and characterization of serogroup M among Nepalese isolates of *Dichelobacternodosus*, the transmitting agent of footrot in small ruminants. *Veterinary Microbiology* 62: 217-33.
3. Hussain I, Wani S A, Qureshi S D, Farooq S. 2009. Serological diversity and virulence determination of *Dichelobacter nodosus* from footrot in India. *Molecular Cell Probes* 23: 112-114
4. Sandhya rani K, Raniprameela D, Sreenivasulu D, Vijayalakshmi S and Karthik A (2016) Development of inactivated whole cell vaccine against ovine footrot. *International journal of veterinary science.*5(2): 103-106.
5. Sreenivasulu D, Vijayalakshmi S, Raniprameela D, Karthik A, Wani S A and Hussaian I 2013 Prevalence of ovine footrot in the tropical climate of

southern India and isolation and characterization of *Dichelobacternodosus*. *Rev. Sci. Tech. Off. Int. Epiz, OIE* 32(3).

6. Vijayalakshmi, S., Sreenivasulu, Raniprameela, D., Vinodkumar, N., Karthik, A. (2020). Phenotypic virulence characterization of Dichelobacternodosus from the cases of ovine footrot in Andhra Pradesh and Telangana states of India. *Int. J. Curr. Microbiol. App. Sci.,* (2020) 9(6)
7. Vinodkumar N, Karthik A, Vijayalakshmi S and Sreenivasulu D 2015 Phylogenetic analysis of *Dichelobacternodosus* serogroup-specific *fimA* gene from ovine footrot in Andhra Pradesh. *Vet World*. 8(5): 567–571.
8. Wani S A, Samanta I, Bhat M A and Buchh A S 2004 Molecular detection and characterization of *Dichelobacternodosus* in ovine footrot in India. *Molecular and cellular probes* 18: 289 – 291.
9. Wani S A, Samanta I and Kawoosa S 2007. Isolation and characterization of *Dichelobacternodosus* from ovine and caprine footrot in Kashmir, India. *Research in Veterinary Science* 83: 141-144.
10. Witcomb LA, Green LE, Kaler J, Ul-Hassan A, Calvo-Bado LA, Medley GF, Grogono-Thomas R, Wellington EM. 2014. A longitudinal study of the role of *Dichelobacternodosus* and *Fusobacterium necrophorum* load in initiation and severity of footrot in sheep. Prev Vet Med. 2014 Jul 1; 115(1-2): 48-55. doi: 10.1016/j.prevetmed.2014.03.004. Epub 2014 Mar 16.

Chapter 18

Approaches in Diagnosis of Diseases of Big Cats

Nidhi Rajput

Assistant Professor, School of Wildlife Forensic and Health, Nanaji Deshmukh Veterinary Science University, Jabalpur – 482 001, Madhya Pradesh
e-mail: nidhi3rajput@gmail.com

"Big cat" is the common name given to the members of the genus *Panthera*, *Acinonyx* as well as few ones from the genus *Felis*. These include tiger, leopard, lion, cheetah, cougar, jaguar *etc.* (Davis and Murphy, 2010). Big cats are widespread across the globe and India is also the host of this group of animals. India is the only country in the world to have all the three prime members of the big cat family *i.e.* Royal Bengal tiger, Asiatic lion and Indian leopard. In addition to these three commonly found species, big cats in India include snow leopard and clouded leopard. Other small and medium-sized wild cats are also found in India such as jungle cat, leopard cat, caracal, lynx *etc.* However, big cats form the major part of apex predators in the forest ecosystem and hence, are considered key to maintaining forest health in general.

Is Disease Really a Threat to Big Cats?

Wild species are facing extinction threats because of habitat destruction as well as poaching and big cats are no exception being one of the important mammalian groups exploited in illegal wildlife trafficking. Although, threats to the conservation of big cats are not limited to these issues but go beyond affecting the health of big cats in their natural habitat also. Wildlife, earlier, used to be considered healthy and diseases were considered the part of natural selection to control and balance the population in the forest ecosystem. However, climate change and habitat destruction made it worse by putting many wild species under threats of extinction resulting from unforeseen epidemics. In addition to these, encroachment of forest areas by human and feral animals provide the opportunity of interactions with wild carnivores especially big cats resulting in disease transmission and amplification (Sidhu *et al.*, 2019 and Nayak *et al.*, 2020). In such an imbalanced scenario, sometimes diseases may not result in an epidemic but in turn, may reduce the general vigour and make the species prone to catastrophic events. These issues raise concern for disease detection in endangered species which requires long-term widespread disease surveillance.

On the other hand, big cats are also maintained in the zoological parks with education and awareness purposes for researchers, students and the general public. Under captivity, big cats may often face stresses which can detriment the condition of the animal if exposed to infectious diseases. As many infectious diseases are contagious, these spread rapidly affecting other carnivores in the vicinity. Preventive measures like vaccination are conducted in captive big cats against the major infectious diseases including feline panleucopaenia, rabies, feline rhinotracheitis but there are many diseases that have emerged in big cats, such as canine distemper, against which vaccination is still not widely practised in Indian zoos. Thus, under captive conditions, it is of utmost importance to promptly diagnose the disease for its timely management.

Timely Retrieval and Proper Sampling

Ante-mortem and post-mortem diagnosis both play important roles in managing infectious diseases.

Captive big cats are often prone to infectious diseases and hence regular screening of the stock is recommended. Disease diagnosis in captive big cats interestingly depends on the routine observations made by the animal keepers who notice any abnormal behaviour at first due to their close interactions with the animals. Although, periodical screening is also conducted in all the zoological parks in which various biological samples are collected and tested in the laboratory for the presence of pathogens.

Contrary to this, in the case of free-range big cats, the ailing animal is seldom observed in deep forested areas and usually, the carcass is retrieved for further examination. Carcass retrieval is also demanding and sometimes degenerated changes make it difficult to diagnose the disease if suspected. Therefore, it is always recommended to screen for the presence of pathogens, in case of any mortality event in big cats. Additionally, samples are also collected opportunistically from the free-range big cats during the immobilization or rescue events for screening the common pathogens and also to create the database for future utilizations. In the field conditions, adequate and proper sampling cannot be ensured which require laboratories to find ways to retrieve the data using an array of diagnostic tests.

General Protocol for Biological Sampling

When there is abnormal behaviour reported in captive big cats then it is comparatively feasible to collect the biological samples by taking the animal into the squeeze cage whereas, in free-range big cats, chemical immobilization is the only option to collect the samples through invasive techniques. This is the reason that in free-range conditions, disease diagnosis in big cats is a complex process. Although, when a carcass is retrieved then every possible biological sample is collected and processed using different laboratory techniques which gives an idea about the health status of the deceased animal as well as other big cats in that particular area.

Following is the protocol adopted by the School of Wildlife Forensic and Health, Jabalpur which aims at collecting the samples to screen out important pathogens of big cats.

Histopathological Investigation

Sl.No.	*Biological Sample*	*Preservative*	*Transportation temperature*
1	Tissue pieces from vital organs such as lungs, heart, liver, kidney, spleen, brain	10 per cent Formalin solution	25-28°C

Disease Investigation

Sl.No.	*Disease/Pathogen*	*Biological Sample*	*Preservative*	*Remarks*
1	Canine distemper	Conjunctival swab	RNA*later* solutionor any RNA stabilizing agent	4°C/Cold chain
		Brain tissue		
		Other organs *i.e.* urinary bladder, lungs, lymph nodes	RNA*later*solution or Phosphate buffered saline (PBS)	4°C/Cold chain
2.	Feline panleucopaenia/Canine parvo virus infection	Rectal swab	Phosphate buffered saline (PBS)	4°C/Cold chain
		A small loop of intestines containing ingesta tied at both the ends		
3.	Leptospirosis	Urine sample	—	4°C/Cold chain
		Kidney tissue	Phosphate buffered saline (PBS)	4°C/Cold chain
4.	Tuberculosis	Tissue sample including the nodular lesions on the lungs	Phosphate buffered saline (PBS)	4°C/Cold chain
		Impression smears prepared from the nodules	Fix using the methanol or heat	25-28°C

In general, two sets of tissue samples in 10 per cent formalin as well as in RNA stabilizing solution are collected. Additionally, swabs from orifices and lesions are collected in an RNA stabilizing agent and then all these samples are transported in a cold chain to the concerned laboratories. While diagnosing the diseases in big cats, sampling plays the most important role as improper sample collection and preservation may create ambiguity in laboratory investigations. Therefore, the approach for disease diagnosis in big cats initiates with the timely retrieval of biological samples, proper preservation and prompt dispatch to the laboratory in ideal transportation conditions.

Conventional v/s Advanced Diagnostics

There is an array of diagnostic tests which is conducted on various biological samples to diagnose the diseases or else to understand the pattern of mortality events in big cats. Conventional methods such as immunological and pathological investigations are routinely practised in laboratories.

- ✰ **Immunological investigations:** In captive facilities, most of the big cats are vaccinated against common pathogens. Still, there might be pathogens in the enclosure or surroundings that can affect the health of associated big cats. Similarly, in free-range conditions, big cats interact with a vast number of species including feral animals hence screening for the antibodies against a range of pathogens gives an idea about the exposure status of the animal. This also indicates the presence of pathogens in that particular habitat. Alternatively, feral animals, as well as domestic counterparts in the peripheral villages of the protected area, are also screened periodically to assess the disease status so that suitable management can be adopted to avoid any disease transmission to the big cats and other wild carnivores (Nayak *et al.*, 2020).
- ✰ **Pathological investigations:** Histopathology is a part of SOP (Standard Operating Procedure) while investigating the mortality in big cats. This includes routine staining with the H and E method which depicts the pattern of death however there are other techniques also such as immunohistochemistry which can detect the presence of pathogens in the tissue samples collected during the post-mortem examination (Beineke *et al.*, 2015). Immunohistochemistry is a highly sensitive and specific method for the detection of antigen in post-mortem samples but has certain limitations for its use in living big cats.

  Conventional diagnostics are often easier to perform in the samples of wildlife as these require less complicated preservation and dispatch of the biological samples. However, when broad and long-term studies including the phylogeny of the infectious organism need to be conducted then in such situations, gene-based diagnosis is considered the precise and relatively quick method.

- **Molecular diagnosis:** A lot of research work has already been conducted on the pathogens in the domestic carnivores and research is also going on continuously in their wild counterparts. Both domestic and wild animals share many common pathogens however considering the viral pathogens, with special reference to RNA viruses, mutations have been observed. Although, gene-based diagnoses performed in domestic animals have also been found to be applicable in the big cats (Duarte *et al.*, 2009; Seimon *et al.*, 2013; Stephenson *et al.*, 2013). Molecular diagnosis is standard and sensitive but proper field sampling and preservation requires quality maintenance to get distinct laboratory results.
- **Next-generation sequencing:** Large scale mortalities in lions at Tanzania's Serengeti National Park in 1994 first attracted the attention of epidemiologists to find the reason behind the epidemic of canine virus in big cats. Although, canine distemper is no more limited to canids and affects a range of carnivore hosts even inclusive of non-human primates. Nikolin *et al.* (2016) analyzed 13 CDV variants obtained between 1993 and 2011 from wild carnivore species and the dog of Serengeti's ecosystem through next-generation sequencing (NGS) technology. They observed novel substitution in the CDV-V protein and rare amino acid combination in the region of the CDV-H protein that binds to SLAM (CD 150) host cell receptors in all the non-canid 'Serengeti' strains during the 1994 epidemic. Thus, such advanced technologies assist in finding the variations in the epidemiology of infectious diseases in novel host species. Several other infectious diseases are emerging in the big cats and in such a scenario; NGS may help in penalizing the pathogens present in a particular sample so that its possible impact can be recognized.

Species Variation and Host Response to Pathogens

Studies of the gene-expression profile in response to pathogens in big cats are going on. This is required especially to understand the species variation in response to novel pathogens as well as to the pathogens which were earlier not reported in the big cats. For example, neurological signs are dominant in canine distemper but only 49 per cent of the reported records of CDV infections in the big cats present clinical signs of the disease (Martinez-Gutierrez and Ruiz-Saenz, 2016). Some animals can acquire the infection, suffer a subclinical and/or asymptomatic disease and remain seropositive for years. Hence, genomic investigations help to understand the entry and specificity of CDV through cellular viral receptors such as SLAM or Nectin-4 in non-native hosts. Therefore, studies on host-response to other pathogens such as variants of coronaviruses are also being conducted to find out the possible spill-over as well as spill-back events that may occur at the domestic-wildlife-human interface (Mallapaty, 2021).

Challenges Ahead

To date, research studies have primarily been conducted on the samples available from captive facilities. Studies on free-range big cats still require a large database to understand the epidemiology, species variation and host response to various diseases which are present or not reported due to the unavailability of relevant biological samples in the field conditions. Disease surveillance is an ongoing process and with the emergence of pathogens in novel hosts, it needs to be accelerated and uniformly applied.

Nevertheless, while studying the disease system in big cats, field sampling is most important and complicated at the same time. Although, training and capacity building of frontline staff both in the forest department as well as in the animal husbandry department on scientific sample collection has been initiated but it would take time to sensitize the field staff on these issues. Hence, meanwhile, laboratories should focus on developing multifaceted testing protocols involving conventional as well as advanced technologies while approaching the disease diagnosis in big cats using verities of biological samples preserved in diverse ways.

References

1. Beineke, A., Baumgärtner, W. and Wohlsein, P. (2015). Cross-species transmission of canine distemper virus—an update. *One Health*, **1**: 49-59.
2. Davis, B.W., Li, G. and Murphy, W.J. (2010). Supermatrix and species tree methods resolve phylogenetic relationships within the big cats, *Panthera* (Carnivora: Felidae). *Molecular Phylogenetics and Evolution*, **56**(1): 64–76.
3. Duarte, M.D., Barros, S.C., Henriques, M., Fernandes, T.L., Bernardino, R., Monteiro, M. and Fevereiro, M. (2009). Fatal infection with feline panleukopenia virus in two captive wild carnivores (*Panthera tigris* and *Panthera leo*). *Journal of Zoo and Wildlife Medicine*, **40**(2): 354-359.
4. Nayak, R.M., Jadav, K.K., Rajput, N., Gupta, S., Rokde, A. and Singh, K.P. (2020). Surveillance of major canine pathogens in feral dogs and big cats at the domestic-wildlife interface in Panna tiger reserve, India. *Journal of Animal Research*, **10**(2): 303-308.
5. Nikolin, V.M., Olarte-Castillo, X.A., Osterrieder, N., Hofer, H., Dubovi, E., Mazzoni, C.J., Brunner, E., Goller, K.V., Fyumagwa, R.D., Moehlman, P.D., Thierer, D. and East, M.L. (2016). Canine distemper virus in the Serengeti ecosystem: molecular adaptation to different carnivore species. *Molecular ecology*, doi: 10.1111/mec.13902.
6. Martinez-Gutierrez, M. and Ruiz-Saenz, J. (2016). Diversity of susceptible hosts in canine distemper virus infection: a systematic review and data synthesis.*Veterinary Research*, **12**: 78.

7. Seimon, T.A., Miquelle, D.G., Chang, T.Y., Newton, A.L., Korotkova, I., Ivanchuk, G., Lyubchenko, E., Tupikov, A., Slabe, E. and McAloosea, D. (2013). Canine distemper virus: an emerging disease in wild endangered Amur tigers (*Panthera tigrisaltaica*). mBio 4(4): doi: 10.1128/mBio.00410-13.

8. Sidhu, N., Borah, J., Shah, S., Rajput, N. and Jadav, K.K. (2019). Is canine distemper virus (CDV) a lurking threat to large carnivores? A case study from Ranthambhore landscape in Rajasthan, India. *Journal of Threatened Taxa*, **11**(9): 14220-14223.

9. Stephenson, N., Swift, P., Moeller, R.B., Worth, J. and Foley, J. (2013). Feline infectious peritonitis in a mountain lion (*Puma concolor*), California, USA. *Journal of Wildlife Diseases*, **49**(2), 2013, pp. 408–412.

Chapter 19

Current Status of Bovine Herpesvirus-1 Infections in India

Vipan Kumar, Rinmuanpuii Ralte, Naripjeet Kaur and P.N. Dwivedi

Department of Veterinary Microbiology, Khalsa College of Veterinary and Animal Sciences, Ram Tirath Road, Amritsar – 143 002, Punjab

Bovine Herpesvirus-1 (BHV-1) belongs to family *Herpesviridae* subfamily *Alphaherpesvirinae*. The virus is responsible for causing Infectious bovine rhinotracheitis (IBR), infectious pustularvulvo vaginitis (IPVV), infectious balanoposthitis and abortions. Generally the abortions occur during the third trimester of pregnancy or during 4-7 months.

Virus Structure

BHV-1 is an enveloped virus and hence it is highly sensitive to organic solvents, UV radiations, temperature, pH variations and detergents. The approximate diameter of the virion is 1560 nm with icosahedral symmetry of the nucleo capsid. The genome of the virus is consisting of single molecule of double stranded DNA with an approximate size of 125 to 235 kbp. The virus replicates in the host cell nucleus and produces intra nuclear inclusion bodies. The virus grows rapidly in cultured cells and lyse the infected cells. The virus establishes the latency in sensory ganglion especially in sciatic and trigeminal ganglia. Herpesvirus can survive from one generation to the next through persistent; often latent infections from which the virus is periodically reactivated and shed. Reactivation of the infection is associated with various stress factors including transportation, adverse weather conditions, overcrowding and intercurrent infections. In general the transmission requires close contact particularly mucosal contact such as coitus, licking and muzzling between mother and the offsprings.

The herpesvirus virions contain more than 33 structural proteins, out of which 14 structural proteins are present in the nucleocapsid, 13 are associated with viral envelope, 2 are associated with the DNA the viral genome and 4 are unclassified.

Virus also possesses 12 different glycoproteins which are located in viral envelope and protrude as the peplomers they are termed as gA, gB, gC....and so on. One of the peplomer protein is gE which possesses the Fc receptor activity and binds the normal IgG.

Sub Types of BHV-1

So far various subtypes of the virus have been reported from various parts of the world. They are;

1. Subtype 1.1
2. Subtype 1.2a
3. Subtype 1.2b
4. Subtype 1.3

Subtype 1.1 is mostly involved in respiratory tract diseases and may also cause abortions. This subtype is always included in vaccines,

Subtype 1.2a causes infectious balanoposthitis/infectious pustular vulvovaginits syndrome and may also cause abortions and mild respiratory tract disease

Subtype 1.2b is implicated in infectious balanoposthitis/infectious pustularvulvovaginits syndrome but it has not been found to be associated with the abortions.

Subtype 1.3 is an encephalitic form of the disease

Virus Cultivation

BHV-1 virus is generally cultivated in homologous host cells. Madlin Darby Bovine Kidney (MDBK) cells are considered as the most appropriate cell lines for the growth of the virus. However, virus has also been cultivated in Vero cells and other cell lines. The virus grows well and the intranuclear inclusion bodies are characteristics of herpesvirus infections. Virus specific proteins are also found in the plasma membrane, where they are involved in cell fusion.

Virus Transmission

Bovine herpes virus is extensively excreted in nasal, ocular and genital discharges (upto 10 to 14 days). Breeding bulls infected with the virus shed it in their semen. Placental cotyledons from the aborted foetus also carry large amount of the virus. Thus a healthy animal when it comes in contact of any one of the above may get the infection. The major route of transmission occurs respiratory or genital secretions. Aerosol transmission is the most efficient way of virus transmission for a short distance contact. In latently infected animals the virus gets activated during stress or adverse environmental conditions and is shed in various secretions and excretions. Infected semen used for insemination purpose spreads the virus into susceptible herds.

Respiratory Form

The respiratory form of the BHV-1 infections is termed as Infectious Bovine Rhinotracheitis (IBR) where morbidity reaches 100 per cent and the mortality upto 10 per cent in complicated cases. The disease may appear as mild, subclinical or severe. The most commonly visible signs are fever, anorexia, profuse nasal discharge, conjunctivitis (unilateral or bilateral) and depression. The mucous membrane in the nasal mucosa becomes hyperemic and red giving the reason to name the disease as "Red nose". Very tiny and small pustules develop in the mucosa of the nostrils. The breathing becomes difficult leading to dyspnea. Excessive salivation and deep bronchial cough are quite common signs. In pregnant animals abortions occur during 4-7 months of gestation. In adult cattle the gastrointestinal symptoms are also visible.

Genital Form

The affected animals show high fever, anorexia, depression, micturition is more frequent and painful. The grossly visible symptoms appear is infectious pustularvulvo vaginitis (IPVV). The vulval labia becomes hyperemic and swollen and there is serous discharge. Many small pustules appear on vestibular mucosae. Generally most of the cases appear to be subclinical and may go unnoticed. In breeding bulls, there is infectious balanoposthitis. These lesions are very severe and painful during the acute stage of the disease and the animal refuses to mount. Due to periodic shedding the semen of such bulls always remain infective.

Current Status of Disease

The outcome of cross breeding programme in India resulted in IBR/IPV as an endemic status. This virus/disease was first reported in India in 1976 in UP. Since then now the disease has been reported from almost every Indian state. Seroprevalence studies by many Indian worker suggests that exotic and cross bred cattle are more susceptible to this virus than indigenous breeds.

Diagnosis

1. Virus isolation can be achieved form the samples such as ocular discharge, nasal discharge, genital discharge in MDBK cell lines where typical syncytia formation and intranuclear inclusions can be seen.

2. Viral antigen can be demonstrated in smears collected from ocular, nasal or genital swabs or in frozen tissue sections from aborted fetuses by applying immunofluorescence technique or ELISA.
3. Virus neutralization test can be applied on paired serum samples to detect active infection.
4. Polymerase chain reaction to amplify viral DNA.
5. Real time PCR is now a days used as alternative of viral isolation.

Control

1. For the control of BHV-1 Infection various kinds of vaccines are available which include inactivated, sub unit and modified live vaccines.
2. Vaccination reduces the severity of clinical signs but cannot prevent the infection. Regular re-vaccination at 6 months interval can reduce risk of infection.
3. Modified live vaccine should not be given to pregnant animals as it causes abortions. Most vaccines are given via intranasal route.
4. Currently marker vaccines are available which facilitates either eradication or reduction of infection in national herds to save the economy.

Chapter 20

Technology Advances in Diagnostic Microbiology

T.G. Sumithra*, K.J. Reshma and V.N. Anusree

Marine Biotechnology Division, ICAR-Central Marine Fisheries Research Institute, Ernakulum North PO, Kochi – 682 018, Kerala
**e-mail: sumithravet@gmail.com*

Introduction

Infectious diseases are disorders caused by various microbes, including bacteria, viruses, fungi, protozoa, and parasites. These are one of the leading causes of morbidity and mortality in both terrestrial and aquatic animals including humans, representing significant threats to the environment, human and animal welfare, and the economy. Rapid and accurate determination of the microbial involvement in a particular case is fundamental to the management of any infectious disease. Accordingly, diagnostic methods and strategies play important role in defining the appropriate treatment, monitoring the therapeutic response, judging the prognosis, epidemiological surveillance programs as well as in the containment strategy of pandemic/panzootic diseases. Diagnostic microbiology is the field in microbiology, which comprises the methods and strategies to rapidly detect and accurately identify the pathogenic microbes in the test specimens. The history of diagnostic microbiology begins with Antonie van Leeuwenhoek, who first observed microbes through his primitive microscope. The works of Louis Pasteur (1822-1895) and Robert Koch (1843-1910), made significant revolutions in this field. The advances in microbiology, biotechnology, and immunology have opened new approaches in diagnostic microbiology, and the rapidly progressing nucleic acid sequencing platforms and bioinformatics promise to continue this trend (Buchan and Ledeboer, 2014).

Diagnostic Microbiology: Principles

The purpose of diagnostic microbiology is to define the infectious nature of the diseases and elucidate the treatment options through detection, quantification, and characterization of the infectious agents. The characterization includes identification, sub-typing, phylogenetic analysis, virulence determination, and drug resistance assessment. The diagnosis of a particular disease is done based on any of the five following criteria; (1) clinical symptoms and history, (2) direct microscopic examination for morphological characterization, (3) *in vitro* or *in vivo* cultivation, (4) measurement of microbe-specific immune responses, and (5) detection of microorganism-specific molecules, especially nucleic acids. As the history and clinical manifestations in many infectious diseases are often nonspecific, it is necessary to apply different laboratory tests based on the last 4 principles to identify the culprit microbes. As veterinary and fish medicine targets terrestrial and aquatic animals of diverse categories, which often have their unique pathogens, veterinary/aquatic animal diagnostic microbiology faces even greater challenges than its human medical counterpart (Liu, 2013).

Diagnostic Tests in Microbiology: Desirable Attributes

The primary requirements of any diagnostic microbiological test are high specificity and sensitivity. The sensitivity of a technique indicates the minimum number of microbes required in the test specimen to produce a positive result in a positive specimen. The specificity is the ability of the test to correctly identify the involved pathogen. Apart from these two primary attributes, the ability to perform within a small sample volume/quantity, rapidity to produce the results, precision, simplicity to perform, and affordability are the desirable attributes of diagnostic microbial tests. As early diagnosis and specific treatment have a crucial role in curing the patient or in reducing their morbidity and mortality of certain diseases, sensitivity, specificity and rapidity are the most desirable attributes. Other desirable collateral properties include the possibility of being automated, the ability to detect multiple pathogens simultaneously, and being cost-effective (Vila *et al.*, 2017). The major commercially available microbial diagnostic/identification kits in India, along with their manufacturing companies are depicted in Table 20.1.

Technology Advances in Morphological Characterization of Microbes

Direct microscopic examination is based on the distinct morphological features *viz.*, size, shape, movements, internal and external components of the pathogen. Application of light microscopy and electron microscopy (both transmission and scanning electron microscopy), together with certain stains/ dyes helps to reveal the morphological details, allowing the identification. Application of highly sensitive fluorescent dyes, specific fluorescent sensor molecules in electron microscopy, fluorescence microscopy, or atomic force microscopy has improved the sensitivity and specificity of the morphological characterization of pathogens especially viruses. Besides the diagnosis, these techniques were utilized to unravel the paradigms of pathogenesis (Liu, 2013).

Table 20.1. Commercially available Microbial Diagnostic and Identification Kits in India

Sl.No.	*Name of Kit*	*Suppliers/ Manufacturers*
1	**Probe PCR kits (PPKs)**	
	Hi-PCR® *Salmonella* PPK, Hi-PCR® *Mycobacterium tuberculosis* PPK, Hi-PCR® Generic *E. coli* PPK, Hi-PCR® *E. coli* O157:H7 PPK, Hi-PCR® *Vibrio cholerae* PPK, Hi-PCR® *Salmonella* Quantification PPK, Hi-PCR® *E. coli* O157:H7 Quantification PPK, Hi-PCR® Methicillin Resistant *Staphylococcus aureus* (MRSA) (Multiplex), Hi-PCR® Vancomycin Resistant *Enterococci* (Multiplex), Generic Dengue detection kit, Hi-PCR® Dengue serotyping PPK, Hi-PCR® malaria PPK, Hi-PCR® *Plasmodium* species (Multiplex) PPK, Chikungunya detection kit, Hi-PCR® *M. tuberculosis* PPK, Hi-PCR® *Leptospira* PPK, Hi-PCR® COVID-19 RUO PPK, Hi-PCR® Generic *E. coli* PPK, Hi-PCR® *E.coli* O157:H7 PPK, Hi-PCR® *V. cholerae* PPK, Hi-PCR® *Legionella* species PPK, Hi-PCR® Human Papilloma Virus Genotyping (Multiplex) PPK, Hi-PCR® Hepatitis A Virus PPK, Hi-PCR® Hepatitis E Virus PPK, Hi-PCR® Rotavirus PPK, Hi-PCR® Enterovirus PPK, Hi-PCR® Astrovirus PPK, Hi-PCR® Extended Spectrum β-Lactamases Gene (Multiplex) PPK, Hi-PCR® Carbapenemase Gene (Multiplex) PPK, Hi-PCR® colistin resistance PPK, Hi-PCR® Extended Spectrum β-lactamases, Gene Quantification PPK, Hi-PCR® colistin resistance encoding gene quantification PPK, Hi-PCR® Carbapenemase gene quantification PPK, Hi-PCR® Leptospira detection kit, Hi-PCR® White Spot Syndrome Virus PPK, Hi-PCR® *Enterocytozoonhepatopenaei* PPK, R and D Systems™ MycoProbe Mycoplasma detection kit	Either Himedia or Thermo scientific
2	**SyBr PCR kits (SPKs)**	
	Hi-PCR® *M. tuberculosis* SPK, Hi-PCR® *E. coli* O157:H7 SPK, Hi-PCR® *Salmonella* SPK, Hi-PCR® *Listeria monocytogenes* SPK, Hi-PCR® *Campylobacter jejuni* SPK, Hi-PCR® *V. cholerae* SPK, Hi-PCR® *Legionella pneumophila* SPK, Hi-PCR® *Klebsiella pneumoniae* SPK, Hi-PCR® *Pseudomonas aeruginosa* SPK, Hi-PCR® *Bacillus subtilis* SPK, Hi-PCR® *Shigella* spp. SPK, Hi-PCR® *S. aureus* SPK, Hi-PCR® *Acinetobacter baumanni*SPK, Hi-PCR® Mycoplasma SPK, Hi-PCR® *Candida albicans* SPK, Hi-PCR® Generic Dengue SPK	Himedia
3	**Conventional PCR detection kits for pathogens**	
	Hi-PCR® Palm *E.coli* 0157:H7 detection kit (DK), Hi-PCR® Palm *Salmonella* DK, Hi-PCR® Palm *M. tuberculosis* DK, Hi-PCR® Palm *L. monocytogens* DK, Hi-PCR® Palm *B. subtilis* DK, Hi-PCR® Palm *C. albicans* DK, Hi-PCR® Palm *Pseudomonas aeruginosa* DK, Hi-PCR® Palm *Staphylococcus aureus* DK, Hi-PCR® Palm Generic *E.coli* DK, Hi-PCR® Palm Diarrhoegeni c *E.coli* DK, Hi-PCR® Palm *Shigella* spp. DK, Hi-PCR® Palm *K. pneumoniae* DK, Hi-PCR® Palm *A. baumanni* DK, Hi-PCR® Palm *Salmonella enterica subsp. enterica* DK, Hi-PCR® Palm MRSA DK, Hi-PCR® Palm Malaria DK, Hi-PCR®	Either Himedia or Thermo scientific

Sl.No.	*Name of Kit*	*Suppliers/ Manufacturers*
	Palm *Clostridiumsporogenes* DK, Hi-PCR® Palm Generic Dengue DK, Hi-PCR® Palm White Spot Syndrome Virus DK, Hi-PCR® Human Papilloma Virus kit, Meridian Bioscience™ Alethia™ Mycoplasma Direct DNA amplification assay	
4	**Pathogenic bacterial detection kits for food and water**	
	Hi-PCR® Generic *E. coli* Food SPK, Hi-PCR® *Salmonella* Food SPK, Hi-PCR® *Listeria monocytogenes* Food SPK, Hi-PCR® *E.coli* O157:H7 and *Salmonella* spp. DK, Hi-PCR® *E.coli* O157:H7 food DK, Hi-PCR® *Salmonella* food DK, Hi-PCR® *Vibrio cholerae* water DK, Hi-PCR® *E.coli* O157:H7 water DK, Hi-PCR® *Salmonella* water DK, Hi-PCR® *Legionella* spp. water DK	Himedia
5	**Other kits**	
	Water Bacteria test kit for swimming water and water tanks, Helico Bacteria Pylori test kit, Remel™ PathoDX™ Strep Grouping kit for *Streptococci* spp., Streptex™ Latex agglutination test (LAT) for *Streptococci* spp., Oxoid™ DrySpot™ *E. coli*Serocheck O103 LAT Kit for *E. coli,* DrySpot™ *E. coli* O157 LAT for *E. coli, Salmonella* Test Kit using LAT, Oxoid™ Oxoid™ *Salmonella* LAT for *Salmonella,* Oxoid™ Microbact™ Staph Fast Blue Reagent Set for *S. aureus,* DrySpot™Staphytect Plus™ LAT for *S. aureus,* BactiStaph™ LAT kit for *S. aureus,* Staphaurex™ Plus LAT for *S. aureus,* Staphylococcal Enterotoxin Reversed Passive LAT Kit, Otsuka America BreathTek™ UBT *H. pylori* kit, H. PYLORI QUIK CHEK™ stool antigen test, Abbott™ Acceava™ *H. pylori* kit, Remel™ Cryptococcus antigen test kit, Latex Meridian Bioscience™ Premier™ Cryptococcal antigen, Wellcolex™ Color Shigella kit, Meridian Bioscience™ ImmunoCard™ STAT Cryptosporidium/Giardia test kit, Arlington Scientific Toxoplasma IgG ELISA test kit, Arlington Scientific Toxoplasma IgM ELISA test kit, Diamedix™ Immunosimplicity™ Toxoplasma IgG test kit, Meridian Bioscience™ ImmunoCard Mycoplasma test kit, Meridian Bioscience™ MeriFluor™ Chlamydia direct fluorescent test kit, QuidelQuickVue™ Chlamydia Test, Remel™ PathoDx™ Chlamydia culture confirmation kit, IMAGEN™ Chlamydia Kit using direct immunofluorescence assay, Abbott™ BinaxNOW™ Legionella urinary antigen test kit, DrySpot™ *L. pneumophila* Serogroup 1 LAT, Legionella LAT, Xpect™ C. difficile Toxin A/B Test, Meridian Bioscience™ Alethia™ C. difficile test tit, Meridian Bioscience™ Revogene™ *C. difficile,* Remel™ BactiCard™ Candida kit, Remel™ *Candida albicans* test kit, Fisher Healthcare™ Sure-Vue™ Rota test kit, Xpect™ Rotavirus test, Meridian Bioscience™ ImmunoCard™ STAT Rotavirus test kit, ProSpecT™, Giardia/ Crytopsoridium microplate assay, ProSpecT™ Giardia microplate assay, Xpect™ Giardia test, Giardia/cryptosporidium QUIK CHEK™, Fisher Healthcare™ Sure-Vue™ Adeno test kit, Quidel™ QuickVue™ Adenoviral conjunctivitis test kit, Fisher HealthCare™ Sure-Vue™ Signature Strep A test kit for *Streptococci* spp.	Angel Chemicals/ Endo View Yi Gripp/Thermo scientific

Technology Advances in Immunological Diagnostic Methods

Specific immune reactions between host antibodies and microbial antigens are known to provide highly sensitive, specific, and rapid identification of microorganisms. For immunologic detection of microbial antigens, latex particle agglutination, immunofluorescence assay, enzyme immune assay, and lateral flow chromatographic immunoassay (LFCIA) are the most frequently used techniques in the field conditions. In comparison to the regular techniques, the LFCIA procedures are easy to perform, interpret, and require very little time. Phage display is recently applied to discover the specific antigenic peptides of the pathogens, which are then synthesized either chemically or through recombinant DNA technologies, and used for the production of specific antibodies. The use of microorganism-specific monoclonal and recombinant antibodies has improved the reagent availability, sensitivity, specificity, and reproducibility of these field diagnostic tests. Another interesting development is the use of chemically synthesized peptides in peptide microarrays as potential diagnostic tools in high-throughput immunoassays. Other new developments in immunological diagnostic techniques include biosensors and nanotechnology (nanoarrays and nanochips). Biosensors involve the use of a microbe-specific antibody and a transducer like electrochemistry, reflectometry, interferometry, resonance, and fluorimetry to convert a biological interaction into a measurable signal. Nanotechnology also offers small-scale platforms to identify an array of infectious agents or serotypes on a single chip (Otto *et al.*, 2014).

Technology Advances in Molecular Diagnostic Methods

Much of the diagnostic microbiology tests has now shifted to molecular-based methods due to the increased sensitivity and specificity and reduced time requirement and labor costs. Progresses in the areas of template preparation, the transition from non-amplified to amplified approaches, detection methods of nucleic acids, sequencing platforms, and bioinformatics have made molecular technologies an obligatory tool in diagnostic microbiology. The recent development of various easy-to-use commercial kits has removed the lengthy protocols and use of hazardous organic solvents in the isolation procedures during the preparation of template nucleic acids from the microbial cultures and clinical specimens. Furthermore, automated, sophisticated, and affordable nucleic acid extraction systems are becoming increasingly available, contributing to the easy performance of template preparation in the diagnostic field. The GeneXpert Dx diagnostic system (Cepheid) and IQuumLiatTM Analyzer are examples of fully automated molecular diagnostic technology technologies. Earlier, the molecular diagnostic procedures chiefly rely on the non-amplified, hybridization signals, and the use of specific probes such as dot blot, Southern blot, and fluorescence in situ hybridization. Technological progress that has happened in this DNA hybridization-based approach is the development of DNA microarray, in which high-density oligonucleotide probes are immobilized on a chip surface, and used to hybridize the complementary sequences (labeled with fluorescent dyes) in a test sample. Subsequently, the generated fluorescence signals are analyzed to identify and quantify the pathogens. The revolutionary invention of the *in vitro* nucleic acid amplification technique namely, polymerase chain reaction (PCR) has significantly simplified molecular diagnostic microbiology. This was followed by different technological advances in the conventional PCR *viz.*, nested PCR, multiplex PCR, real-time PCR (RT-PCR), reverse transcription PCR, random amplified polymorphic DNA -PCR, degenerate oligonucleotide-primed PCR, sequence-independent single primer amplification, helicase dependent amplification, digital PCR and rolling circle amplification to meet the various desirable attributes of diagnostic tests. Apart from PCR, other advanced nucleic acid amplification technologies are also being developed such as nucleic acid sequence-based amplification, ligase chain reaction, strand displacement amplification, Q-β replicase mediated amplification, linear-linked amplification, and loop-mediated isothermal amplification assays (Buchan and Ledeboer, 2014). Another area in which lots of progress are on the detection methods of amplified nucleic acid products. Ethidium bromide and SYBR Green are the commonly used conventional stains. Later, specifically designed probes such as hydrolysis dual-labeled probes (TaqMan®), hybridization probes (LightCycler), molecular beacons, peptide nucleic acid (PNA) probes, TaqMan minor groove binding (MGB™) probes, locked nucleic acid (LNA®) primers, and probes, and scorpions™ are developed. Single strand conformational polymorphism analysis (SSCP), denaturing gradient gel electrophoresis and temperature gradient gel electrophoresis (TGGE) are the major advances in nucleic acid separation techniques. SSCP in combination with capillary electrophoresis provides an automated system for rapid separation of nucleic acid products (Liu, 2013). The development of biochips, which are small analytical devices designed for nucleic acid-based electrical/optic detection (fluorescence or chemiluminescence) is another progress in molecular diagnosis. Nevertheless, DNA sequence analysis is the most accurate method for microbial identification. The development of next-generation sequencing platforms such as Illumina, PacBio, and Nanopore technologies are the major advances in this field, application of these methods has dramatically increased the sequencing capacity and reduced cost per base.

Rapid and Sensitive Technologies for Bacterial Identification and Antimicrobial Profiling

Traditional approaches to the identification and drug susceptibility testing rely on cultivation-based approaches, which can take 2 to 7 days for producing the results, depending on the species of bacteria. While the molecular techniques can speed up the infectious disease diagnosis, a rough estimation of the pathogen/resistance that can be involved in the case is needed, while an unknown or potentially new or broad category of both pathogens and resistance profiles cannot be surveyed. Therefore, next-generation technologies that can speed the direct phenotypic testing of bacteria and monitor the effects of antimicrobial therapies are under development. Ultrasensitive detection methods based on a variety of electrophoretic, electrochemical, and mass spectrometry-based systems have been developed for rapid pathogen identification. Of these, the development of MALDI-TOF (matrix-assisted laser desorption ionization time-of-flight mass spectrometry) devices was the most striking revolution in the routine identification of microbes in diagnostic microbiology laboratories. Application of MALDI-TOF has the potential for the accurate identification of the most Gram-positive and Gram-negative bacteria and many filamentous fungi and dermatophytes at the species level after the standardization, except a few difficult strains that require more attention and further development of the method (Croxatto *et al.*, 2012). VITEK® MS microbial identification system is an example of automated MALDI-TOF technology-based microbial identification. Recent progress in single-cell monitoring is now being utilized for making new solutions that can speed up the bacterial analysis by monitoring changes in growth on a cell-by-cell level. Here, single-cell microbioreactors and microscale incubatorsare being developed to cultivate bacteria in very small volumes, and new types of single-cell analysis methodsare applied to rapidly assess the bacterial viability. These types of methods thus can reduce the turnaround time efficiently, as very small numbers of bacteria can be analyzed (Kelley, 2017). A high throughput approach based on microscopy is recently being tested that tracks the morphology of bacterial cells when they are exposed to different antimicrobial agents (Choi *et al.*, 2014). The electrochemical readout is also applied to speed up the antimicrobial phenotypic testing within miniature testing cells. An application of this technology is done by Besant *et al.* (2015), who explored the dye resazurin to monitor the antimicrobial response of the bacteria trapped in nanoliter wells within a microfabricated device.

Conclusion

Advances in molecular and cell biology and genetics have revealed new technologies for the diagnosis of different infectious diseases. Technologies have significantly improved the speed, sensitivity, and specificity of the diagnostic methods. Further, technological advances have created many fully automated devices for the busy clinical microbiology laboratory. Despite these increasing technologies, development of the rapid diagnostic tests, and the availability of molecular and proteomics-based tests, widespread use of cost-effective, validated commercial assays has yet to be applied in many diagnostic laboratories particularly in developing countries. Nevertheless, technological progress will continue and leads diagnostic microbiology certainly to be a modern discipline, to face inevitable microbial challenges in the future.

References

1. Besant JD, Sargent EH, Kelley SO. (2015). Rapid electrochemical phenotypic profiling of antibiotic-resistant bacteria. Lab Chip, 15, 2799-2807.

2. Buchan BW, Ledeboer NA (2014). Emerging technologies for the clinical microbiology laboratory. Clinical Microbiology Reviews 27 (4), 783-822.
3. Choi J, Yoo J, Lee M, Kim E-G, Lee JS, Lee S, Joo S, Song SH, Kim E-C, Lee JC, Kim HC, Jung Y-G, Kwon S (2014). A rapid antimicrobial susceptibility test based on single-cell morphological analysis. Science Transitional Medicine, 6(267), 174
4. Croxatto A, Prod'hom, G, Greub G (2012). Applications of MALDI-TOF mass spectrometry in clinical diagnostic microbiology. FEMS Microbiology Reviews 36, 380-407.
5. Kelley SO (2017). New technologies for rapid bacterial identification and antibiotic resistance profiling. SLAS Technology 22(2) 113-121.
6. Liu D (2013). Technical advances in veterinary diagnostic microbiology. In: Advanced Techniques in Diagnostic Microbiology. YW Tang and CW Stratton (eds.) Springer publishers, New York. pp: 647-659.
7. Otto C, DH Persing, YW Tang (2014). Diagnostic microbiology Reference Module in Biomedical Research http: //dx.doi.org/10.1016/B978-0-12-801238-3.02392-81
8. Vila J, Gómez, María D, Miguel S, Jordi B (2017). Methods of rapid diagnosis in clinical microbiology: Clinical needs. EnfermedadesInfecciosas y MicrobiologíaClínica 35 (1), 41-46.

Chapter 21

Heavy Metal Toxicity in Animals

Amita Dubey, Arya Anil and Yamini Verma

Department of Veterinary Pathology, College of Veterinary Science and AH, NDVSU, Jabalpur, Madhya Pradesh

Heavy metals are metallic elements which have a high atomic weight and a specific density of more than 5 g/cm3, much greater (at least 5 times) than water. These metals are quintessential to maintain various biochemical and physiological functions in living organisms at very low concentrations; however they become noxious when they exceed certain threshold concentrations (Wallace *et al.*, 1977).

In recent years, with the development of the global economy, both the type and content of heavy metals in the soil have gradually increased, resulting in deterioration of the environment (Sayadi and Rezaei, 2014). The anthropogenic activities like mining, industrialization, automobile emission, waste from industries, agricultural waste, fertilizers, household sewage, ammunition *etc.* are the reasons for the heavy metal pollution in the environment (Kumar *et al.*, 2008). Soils, whether in urban or agricultural areas represent a major sink for metals released into the environment from a wide variety of anthropogenic sources (Singh and Kumar, 2006). Once the soil suffers from heavy metal contamination, it is difficult to be remediated (Chao *et al.*, 2014). In last five decades there is many fold increase in heavy metal concentration in the environment hydrosphere and lithosphere (WHO, 2001).

There are more than 20 heavy metals, among which, four are of particular concern: lead (Pb), cadmium (Cd), mercury (Hg), and inorganic arsenic (As), according to the U.S. Agency for Toxic Substances and Disease Registry. These are among the top six hazards, present in toxic waste sites, especially to human and animals and can cause damaging effects even at very low concentrations (ATSDR, 2000).

Theses heavy metals cause deterimental effects on various biological systems like plants, fishes, animals and human beings (Govind and Madhuri, 2014). Heavy metals enter into the animal body via inhalation, ingestion, and skin absorption. In general, heavy metals are the systemic toxins and produce specific toxic effect on various system like nervous, hepatic, renal, digestive system. (Jarup, 2003).

Heavy metals- lead, arsenic, mercury and cadmium are detrimental to animal as they have long half life, cumulative effect and bio-accumulation in the tissues and alter the enzymatic and metabolic processes. These metals are also potent carcinogens and mutagens. The high accumulation of Hg and Cd caused *minamata*and *itai-itai*diseases, respectively (Wilkinson *et al.*, 2003).

Chronic exposure to sulfhydryl reactive metals results in higher accumulation in tissues, bones, hair and blood. It appears to be highly imperative to monitor these toxic metals in biological materials which indicate the status of environmental degradation (Patra *et al.*, 2007).

Considering the long term detrimental effect of these heavy metals on animal populations, the present seminar is presented with the review on the toxicopathologic aspect of important heavy metals.

Arsenic (As)

Arsenic is a ubiquitous metal, combines with many elements such as oxygen, hydrogen, sulfur, nickel, copper, and lead. Since ancient time it is used in medicine, agriculture, livestock, electronics, and metallurgy.

Common Sources

Contaminated soil and ground water, wood preservative, pesticide, herbicide, fungicide and arsenic dip (Sodium Arsenate), spray of ectoparasiticide in animals, coal and superphosphate and fish meal (Klaassen, 1996).

Areas Affected

Ground water contamination with arsenic was reported from different parts of world, like Vietnam, Bangladesh, Argentina, Japan, Thailand and Hungary, U.S., Southwest Taiwan and Chile (Tsai *et al.*, 1999). In India, West Bengal, Punjab, Haryana, Himachal Pradesh and Uttar Pradesh have high concentration of arsenic in ground water. Arsenic in natural waters generally range between 1 and 2 µg/l, although half of the 4 million tube wells drawing drinking water in Bangladesh exceeds the arsenic standard of 50 ppb (USAID, 1997). In the more contaminated areas, arsenic levels routinely exceed 500 ppb (WHO, 2001). The WHO guidelines describe safety limit of arsenic at 10µg/l and a maximum permissible limit of 50µg/l in drinking water (Steinmaus *et al.*, 2005).

Non ruminants are more susceptible to arsenic toxicity as compared to ruminants. Cat is the most susceptible species, followed by sheep, horse, cattle, pig and birds. The lethal oral dose of sodium arsenite in most species is from 1-25 mg/kg and cats are more sensitive. Toxic range of arsenic after ingestion in cattle is wide with an LD_{50} of 1-25mg/kg for trivalent arsenic and 30-100mg/kg for pentavalent arsenic (Datta *et al.*, 2010).

The toxic dose for oral sodium arsenite (NaAsO2) is 7.5 mg/kg BW in cattle, 11 mg/kg BW in sheep, while for arsenic trioxide (As2O3) is 33-55 mg/kg BW in horses, cattle and sheep (Blood *et al.*, 1992).

Pathogenesis

Inorganic arsenicals are more toxic than the organic forms, and among that trivalent form is 2-10 times higher toxic than the pentavalent form due its oxidation state and solubility (Goyer, 2001).

After accidental exposure of the arsenic either through oral or inhalation route, it gets absorbed easily by the intestinal and respiratory mucosa. Animals are able to tolerate low levels of arsenic; the normal level in cattle tissues is < 0.5 ppm (Monies, 1999). Arsenic levels of > 10-15 ppm in liver, accompanied by clinical signs are considered diagnostic of acute arsenic toxicosis (Fletcher, 1986). Arsenic has affinity for the sulphydryl enzymes of tissue and binds with formation of strong carbon bond (Ercal*et al.*, 2001), and get accumulated rapidly in the liver, kidney, heart and lungs. Metabolized by methylation to monomethyl arsenic acid (MAA) and dimethyl arsenic acid (DAA) it is then excreted through urine (90 per cent) (Yoshida *et al.*, 1998). Clearance of arsenic from most of the species of animals follows two or three phase exponential curve. In first phase >90 per cent, maximum amount is excreted. The half life is 1-2 hrs, 30 hrs, 200 hrs respectively (Neiger, 2001).

Higher affinity is shown to the tissues with high sulphydryl content, specially the keratin, where arsenic accumulates for longer period even up to several years, *i.e.* in hairs and nails (Nandi *et al.*, 2005).

Pathological changes caused by the arsenic is due to its inhibiting effect on Lipoic acid, a tissue respiratory enzyme co-factor (Shila*et al.*, 2005), which causes the oxidative stress. The rapidly dividing the cells of intestinal epithelium, epidermis, kidney, liver and lungs are mainly affected (Calatayud*et al.*, 2014). Capillary vasodilatation leads to submucosal congestion and oedema of intestine in birds and mammals (Khan *et al.*, 2014). Dogs with thiacetarsamide show direct effects on arteries and coagulation system, (Maksimowich*et al.*1997). Placental transfer of arsenic also has been reported in monkeys (Lindgren *et al.*, 1982).

Inorganic arsenic or its metabolites alter gene expression and interfere with cell signals transduction (Kumagai and Sumi, 2007). Demethylation and axonal degeneration produced by organic arsenicals may be due to interference with thiamine requirements for the maintenance of the peripheral nerves (Vahidnia*et al.*, 2008).

Diagnosis

1. Arsenic level can be measured in blood, urine, hair, nails. Blood may not be very useful (Dupler, 2001)
2. Abdominal X-ray can reveal the metallic fragments (Ferner, 2001).
3. Chronic exposure in dogs increases the serum aspartate, AST< ALT level.
4. Urine may contain protein, blood and cast (Osweiler *et al*, 1985). Normal level is less than 1.0 ppm. Increased urine level is 2-100 ppm, and more than 10 ppm in kidney and liver (normal level is 0.5 ppm).
5. Nervous tissue can also be used.
6. Silver diethyl dithiocarbamate spectrophotometric method (ISO 6595:1982) is available for the determination of arsenic; the detection limit is about 1 µg/l.
7. Graphite furnace AAS, hydride generation AAS and ICP-MS are more sensitive. HPLC in combination with ICP-MS can also be used to determine various arsenic species.

Lead (Pb)

Environmental levels of lead have increased more than 1,000-fold over the past three centuries as a result of human activity between the 1950 and 2000 (Callender and Rice, 2000).

Sources

Ore, Galena also called lead glance, the natural form of lead, leaching out in the soil and water, effluents released from batteries, tetra alkyl additives, gasoline, municipal waste incinerator, emission from the pigment and dye industries (Dey *et al*, 1999), vehicle emission, domestic waste, urban sewage contributes 11.6 to 14.45 µg/g (Shrivastav*et al*, 1997), lead arsenate and lead arsenite astringent and ectoparasiticides, lead shot for gun, wine cork covers, old leaded pipes also contribute to the toxicity in domestic and zoo animals. Paint, (lead oxide, lead carbonate, lead sulphate),lead objects- toys, sinkers, drapery, printer dye, news paper, discarded waste batteries, dump oil, oil paint containers, bone-fire ash, greases are all sources of lead for animals.

Whole blood levels of lead in normal ruminants are usually below 0.05–0.25mg/kg; poisoned animals usually have levels above 0.35mg/kg and deaths begin at 1 mg/kg (Dey *et al.*, 1996). Animal reared around the industrial area and highways have higher blood over the minimum toxic level (0.25ppm) attributed to the poisoning. Butler and coworkers (1957) recorded cases of lead poisoning in lambs gazing around the old mining area, Sindhu *et al.* (1994) reported an outbreak of plumbism in cattle and buffaloes in Punjab due to serious contamination of fodder by effluents from a secondary lead smelters. Lead poisioning has been one of the causes of death in vultures in India (Gangoso*et al.*,2009). Concentrations in drinking-water are generally below 5 µg/l, although much higher concentrations (above 100 µg/l) have been measured where lead fittings are present (Edwards *et al.*, 2009). Lead concentrations can also vary according to the period in which the water has been in contact with the lead-containing materials. (WHO, 2011)

Levine *et al*, 1976 reported lead poisioning outbreaks due to environmental contamination at scarp smelter and recorded blood lead level of 945 ug/dl in dogs, 26 ug/dl in horses, 71 ug/dl in cats and 102 ug/dl in cattle.

Pathogenesis

Toxicity of lead occurs due to accidental exposure through oral, inhalation and dermal route. Though oral bioavailabity is low (1-2 per cent).

Lead toxicity has been recorded in fishes, birds, dogs, goats, sheep, buffalo, cow *etc.* (Vellinger *et al.*, 2012) and is the commonest cause of poisioning in farm animals, particularly cattle and young ones are more susceptible. Horses are more sensitive than cattle, sheep and cat, whereas swine, goat and chicken are comparatively tolerant (Siddiqui and Rajurkar, 2008).

Lead once ingested, may settle in reticulum allowing slow absorption for longer period (Siddiqui and Rajurkar, 2008). After absorption, > 90 per cent of lead is bound to hemoglobin molecule of erythrocytes and free form is distributed to various tissues. Followed by it, lead is mainly stored in the bones, teeth and hairs, from where it further comes into circulation (Gulson and Salome, 1995).

Lead exerts cumulative adverse effect by multiple toxic effects on various tissues. It mainly affects the gastrointestinal, haematopoietic, renal, hepatic and nervous system (Needleman and Bellinger, 1991) Lead interferes with the process of haeme synthesis by depressing the delta aminolevulinic acid dehydratase (ALAD) enzyme and thus the incorporation of iron and produces iron deficiency anemia in animals. Lead interferes with the function of thiol group of enzymes and replaces the zinc in some enzymes (Kwong*et al.*, 2004).

Excess lead in kidney damages the glomeruli and tubules causing the hyper filtration, vasoconstriction and increased urate secretion and disruption of sodium potassium pump leading to nephropathy (Diamond, 2005).

Lead competes with the calcium ions and displaces the calcium from bones and calcium binding sites. It adversely affects the bone development process. A recent study in mice reported that lead delays fracture healing at environmentally relevant doses and causes progression of endochondrial ossification (Carmouche *et al.*, 2005). Lead disrupts the modulation of intracellular calcium by 1, 25-dihydroxy vitamin D and D-3 stimulated synthesis of osteocalcin which play a major role in normal bone mineralization (Long and Rosen 1994).

In nervous system lead lowers the GABA concentration which may lead to cerebral edema and neuronal damage centrally and demyelination and reduced nerve conduction velocity peripherally (Gwaltney, 2002). Release of acetylcholine and calmodulin are due to reduced extra cellular calcium. Inhibition of tryptophan in plasma leads to elevation of brain serotonergic pathways (Abadin and Liados, 1999).

Furthermore, lead alters the functioning of the enzyme protein kinase C, a virtually ubiquitous protein important for numerous physiological functions (Murakami *et al.*, 1993).

Diagnosis

1. Lead concentration in blood, serum or plasma.
2. Increased D- ALA in urine
3. Tissue lead level –liver, kidney, brain. Lead at 15-20 ppm in liver consider to be toxic.
4. X-ray of long bones to see the lead lines.
5. Gross and histopathological lesions
6. Differential diagnosis- disease rabies, CD, listeriosis, hypomagnesemia,

Cadmium

Cadmium (Cd) is a biologically essential, naturally occurring and highly active element found in trace quantities throughout the environment. Environmental protection Agency (EPA) has set the safe limit of cadmium in drinking water upto 5 ppb or 0.005mg/L for short term exposures (ATSDR, 1989).

Occurrence of Itai-Itai disease in Fuchi, Japan after world war was due to the contamination of local rice fields by cadmium effluent from lead-zinc processing plant (Yoshida *et al.*, 1999)

The contamination of cadmium in ecosystem lead to death of fishes and small wild mammals. (Rashed, 2001)

Sources

Industry – cadmium is the byproduct of mining and smelting of lead and zinc (Pilgrim and Hughes, 1994), plastic production, electroplating and manufacture of alloys and batteries. Fossil fuels also contribute to environmental contamination.

The dust and waste water from smelting and refining phosphate rocks, fertilizers, coal and oils also discharges nearly 30,000 tones cadmium into environment per year. Recently, cadmium has been detected at high levels in some phosphate and zinc sulfate fertilizers (Fishbein, 1981).

Cadmium concentration is high in sludge irrigated area and the metal is retained in soils as it forms super-phosphate and is then available to plants which may be available to animals that graze on grasses growing in these areas (Simmons *et al.*, 2005).

Shellfish act as reservoirs of cadmium and feed concentrate containing animal tissues like liver and kidney also have high concentration of cadmium (Roberts and Clarke, 1988).

Various cadmium compounds like cadmium oxide and cadmium anthranilate are used as anthelmintics for the treatment of ascariasis in pigs. Cadmium sulphide is used in antiseborrhoeic shampoos (Salvador *et al.*, 2000) and cadmium chloride as feed supplement (Smith *et al.*, 1991).

Pathogenesis

Cadmium enter in animal body either through ingestion or inhalation route, Though the absorption of cadmium from the gastrointestinal tract is a saturable process and is only 1.5 per cent (Nordberg *et al.*, 1985). Diet deficient in Ca++, Fe++, protein or high fat content significantly increases cadmium absorption (Goyer, 1991).

Upon absorption cadmium is transported to different tissues by binding with plasma albumin and red blood cells (Goyer, 1991). The target organs for cadmium toxicity are liver, kidneys, lungs, brain, placenta and bones (Roberts, 1999).

Major part of cadmium is accumulated in the liver, and glutathione conjugated cadmium is excreted in bile. In liver, it induces synthesis of metallothionin, metallothionein-bound cadmium appears in the plasma of exposed animals. Plasma half-life of cadmium ranges between 10-30 years in humans. This Absorbed cadmium is very slowly excreted with approximately equal excretion through urine and faeces (Gregus and Klaassen, 1986). Other routes like milk and sweat also contribute to the excretion of cadmium (Sears *et al.*, 2012).

Neurotoxic action of cadmium chloride is mainly due to inhibition of inositol phosphates which is a necessary signaling pathway for various neurotransmitters, including the action of acetylcholine on muscarinic receptors, noradrenaline on alpha receptors and glutamate on AMPA/KA receptors in oligodendrocytes progenitors (Gusovsky *et al.*, 1987)

Mercury

Mercury is a dense, volatile and mobile liquid at room temperature having a low melting point. According to the International Chemical Safety Council of United Nations, organic form of mercury i.e, Methyl mercury, is one of the most serious pollution threats to the plants and animals. A study estimated that the present global mercury emission into the atmosphere is 5000 tonnes per year, of which 80 per cent isof anthropogenic origin (UNEP, 2002). About 200 metric tonnesof toxic mercury is being out poured as industrial effluent every year in India, resulted the drastic increase in the levels of mercury in rivers, coastal water, soil and food materials, above the acceptable limit (Mukherjee and Zevenhoven, 2006).

Common Sources

Mercury exists in the environment in elemental, organic and inorganic forms. In air through earth crust's, volcanic eruptions and forest fires. Based on data from active volcanoes, it is estimated that the time-averaged volcanic Hg emission is <"700 Mg/yr, or 20–40 per cent of total natural emissions (Pyle and Mather, 2003).

Industrial uses – as manufacture of lamps, barometers and thermometers. Gold mining, production of caustic soda and chlorines.

Inorganic mercurial salts such as mercuric and mercurous salts are also used as preservatives and fixatives, whereas organic mercurials are used as fungicides (Aiello and Mays, 1998). Biotransformed methyl-mercury reaches to humans and animals through the food chain

Pathogenesis

Inorganic mercurial salts are absorbed to a limited extent(7-15 per cent) through gastrointestinal tract although sufficient enough to pose a risk for toxicity(Roberts, 1999). Inorganic mercury in the environment can be methylated by microorganisms to organic form *i.e.*, methyl-mercury, which is readily absorbed in the gastrointestinal tract(90-100 per cent) due to its high lipid solubility and becomes far more toxic when ingested(Berlin *et al.*,2007). The methyl mercury gets biomagnified in plants and animal tissues at the top of the food chain.

Absorbed mercury enter an oxidation reduction cycle in liver and gets oxidized to divalent inorganic cation in red blood cells and lungs of animals and humans. Absorbed mercury is transported primarily in association with the glutathione rich erythrocytes which contain 20 times more mercury than plasma, in exposed animals. Experimental data indicate that kidneys generally contain the highest mercury concentrations, usually at approximately 100-fold the levels found in other tissues including liver or muscle (Clarkson, 1992). All forms of mercury easily cross through placental barrier.

Mercuric ions produce the detrimental effects in cells and orgenelles due to inactivation of various enzymes and oxidative stress and mitochondrial dysfunction.

Since it's binds with the thiol, or sulfdryl groups of proteins in both extracellular and intracellular membranes and organelles, alter the structural proteins and transport process of the cells (Bulger, 1986). It alters the permeability of cell membrane by the formation of mercaptides (sahaphong and trump 1971). Methyl Mercury mimic the essential aminoacid methionine and resulting in faulty protein synthesis (Aschner and Clarkson, 1988).

Mercuric chloride caused the depolarization of the mitochondrial inner membrane which are coupled with Hg++ mediated glutathione depletion and pyridine nucleotide oxidation and altered calcium homeostasis creating an oxidant stress condition characterized by increased susceptibility of the mitochondrial membrane to iron-dependent lipid peroxidation (Lund *et al*, 1993).

Neurons are highly sensitive to mercury either because of their low endogenous glutathione content or in their inefficient glutathione redox activity. Organic alkyl mercurial has been proposed increase in the formation of reactive oxygen species in several brain areas has been observed following intraperitoneal administration of methyl mercuric chloride to rodents (Ali *et al.*, 1992). Inhibition of protein synthesis has been reported in neurons from rats exposed to methyl mercury (Philbert *et al.*, 2000)

Methyl mercury also forms a complex in plasma with the amino acid cysteine, which is structurally similar to the essential amino acid methionine and is incorporated into proteins resulting in faulty protein synthesis (Aschner and Clarkson, 1988).

Chapter 22

Antibiotic Resistance Patterns of Methicillin-Resistant *Staphylococcus aureus*

***B.B. Javia*[1] *and V.R. Nimavat*[2]**

[1]*Associate Professor and Head,* [2]*Assistant Professor,*
Department of Veterinary Microbiology, Veterinary College, KU, Junagadh, Gujarat

Methicillin-resistant *Staphylococcus aureus* (MRSA), a type of bacteria that is resistant to several antibiotics. It is one of the staphphylococcal infection which is difficult to treat due to resistance to several available antibiotics. It is bacteria that are genetically differ from other strains of *Staphylococcus aureus*. The development of MRSA is by natural selection or may be acquired through horizontal gene transfer and is responsible for multiple drug resistance to beta-lactam antibiotics. Beta-lactam (β-lactam) are a broad-spectrum group of antibiotics mainly few penams (penicillin *e.g.* methicillin and oxacillin) and cephems such as the cephalosporins[1].

In 1961, the first known MRSA isolates were reported in a British study, and from 1961 to 1967, infrequent hospital outbreaks occurred in Western Europe and Australia, with methicillin then being licensed in England to treat resistant infections. Other reports of MRSA began to be described in the 1970s. Resistance to other antibiotics was documented in some strains of *S. aureus*. Vancomycin was the only antibiotic used againstit but in 1996, vancomycin resistance was reported in Japan[1]. Few reports suggest in past that MRSA infected patients require more time of hospitalization and high cost compare to not MRSA infection.

In livestock production industry *S. aureus* strains are more dangerous to dairy animals and are of greatest attention for scientific community worldwide as they possess properties like capabilities to produce enzymes, toxins and intrinsic virulence factors which cause invasion and tissue damage to mammary gland. It has also having ability to survive in the keratin of the teat canal of healthy cow and capability to resist phagocytosis (protein-A). Apart from this there are low level of complements and opsonizing antibodies specific to *S. aureus* in milk and most importantly having an ability to resist antibiotic therapy due to production of beta lactamases a group of enzymes that inactivate penicillin and closely related antibiotics[2]. This MRSA strain in intra-mammary dissemination often produce incurable severe intra-herd infections[4]. MRSA strains have been observed to be multi-drug resistant, such as aminoglycosides, macrolides, lincosamides, streptogramins, tetracyclines, *etc.*, which are often used in the treatment of mastitis in veterinary practice[5]. MRSA from bovine sources was first reported in 1975[6]. The presence of MRSA in bovine milk and dairy environments poses potential risk to farm workers, veterinarians, consumers and other farm animals which can have adverse and deleterious public health concerns[7].

Microbiological and Epidemiological Aspect of MRSA

Similar to *S. aureus* methicillin-resistant *S. aureus* is a gram-positive, spherical cocci about 1 micron in diameter. It does not produce spores and is not motile. Morphologically under microscope it appear in grape-like clusters or chains. Compare to other susceptible *S. aureus*, MRSA is slow-growing on a different media and has been found to exist in mixed colonies of methicillin-sensitive *S. aureus*(MSSA). The widely follow method for identification of methicillin-resistant *Staphylococcus aureus* in samples from various biological and environment samples (*e.g.*, nasal, skin, and dust swabs) consists of a pre-enrichment step followed by a selective enrichment step before plating on chromogenic MRSA-selective and -indicative agar. MRSA colonies on blood agar are greyish or yellowish and usually surrounded by a zone of haemolysis. The catalase test can be used to distinguish staphylococci from enterococci. Earlier, oxacillin was used as the marker antibiotic to detect MRSA; however, CLSI now recommends cefoxitin, as it is a good inducer of *mecA* and *mecC* than oxacillin and results in a clear identifiable phenotype. Currently DNA-based methods for direct MRSA detection from clinical samples are used to detect the various genes present in it for final confirmation[8].

Epidemiologically, MRSA responsible for hospital outbreaks is known as health-care-associated MRSA (HAMRSA). Later significant difference in MRSA epidemiology was seen when it was identified in person without previous health-care contact. It was then designated as community-associated MRSA (CAMRSA). In the community MRSA generally causes skin infections. Sometimes, it causes pneumonia and other infections. Untreated cases may converted to sepsis condition. Antibiotic use in livestock treatment increases the risk that MRSA will develop among the livestock. The MRSA strains also have been isolated from persons who are associated with livestock exposure which is known as livestock-associated MRSA (LAMRSA). Generally, animals does not produce symptoms and remain asymptomatic. Companion animals are susceptible to MRSA infection by transmission from their owners and vice versa to humans. There are reports of LAMRSA in individuals with no connection to livestock, and in these cases, spread via environmental contamination or, less commonly, foodborne transmission has been postulated[8].

Various researcher have worked on mastitis in dairy animals with targeting MRSA as etiological agent and also studied its antimicrobial resistance patterns. Moon *et al.* analyzed 3,047 bovine mastitic milk samples from 153 dairy farms in the Republic of Korea, collected from 1997 to 2004. 21 (2.5 per cent of 840) *S. aureus* and 19 (2.4 per cent) coagulase-negative staphylococci were resistant to methicillin[3]. They screened methicillin resistance by disk diffusion test with an oxacillin disk and confirmed by MIC test with oxacillin. Phenotypic methicillin resistance was defined with an oxacillin MIC ≥ 4 µg/mL. They observed 70 per cent of the isolates belonged to four predominant coagulase genotypes (I, II, VII and VIII). Vanderghaeghen*et al* found nearly 10 per cent of the *S. aureus* isolated from bovine subclinical and clinical mastitis in their study were MRSA, a much higher prevalence than previously published. Strains proven to be MRSA were tested for susceptibility to non-β-lactam antimicrobial agents by using the disk diffusion method. The reported strains were ST398, spa types t011 or t567, and had *SCCmec*-type IVa or V[9]. Kumar *et al.* found a 13.1 per cent MRSA prevalence when analyzing 107 strains of *S. aureus* from 195 mastitic milk samples in India[10].

Risk Factors Associated with MRSA Colonization or Infection

There are various risk factors associated with colonization or infection with MRSA in human as well as in livestock. In human person with indwelling implants, drains, and catheters or who share the equipment and having close contact. Old age people with weak immune system or concurrent medical condition play a very vital role in set up of MRSA infection. Individual engaged with livestock rearing and production, veterinarians and paravets, livestock handlers, pet owners are also at risk. Contaminated or untreated milk or milk products may also a source of MRSA to human being.

Mechanism in Development of MRSA

MRSA has emerged globally as a significant nosocomial pathogen of public health concern and it is constantly evolving to resist the antibiotics both in human and veterinary medicine. MRSA isolates are frequently multidrug resistant (MDR), which can result in higher costs and longer treatment times[11].

Antimicrobial resistance is based on genetics which is controlled by the acquisition of extra-chromosomal genetic elements containing genes that are responsible for resistance to certain antibiotics. It include plasmids, transposable genetic elements, and genomic islands. These can be transferred between bacteria through horizontal gene transfer[12]. A most important characteristic of MRSA is its capacity to succeed in the presence of penicillin-like antibiotics, which normally does not allow bacterial growth by inhibiting synthesis of cell wall material. This is due to a resistance gene, *mecA*, which stops β-lactam antibiotics from inactivating the enzymes (transpeptidases) which is primarily require for cell wall synthesis[13]. *mecA* is a gene responsible for resistance to methicillin and other β-lactam antibiotics. After entry of *mecA*, the gene must be integrated and localized in the *S. aureus* chromosome[14].The *mecA* gene, which is responsible for resistance to a number of antibiotics, is mostly present in MRSA and usually absent in MSSA; however, in some cases, the *mecA* gene is present in MSSA but is not expressed[8]. *mecA* encodes penicillin-binding protein 2a (PBP2a), which is different from other penicillin-binding proteins so it does not bind methicillin or other β-lactam antibiotics[14].

Staphylococcal cassette chromosome *mec* (SCC*mec*) is a genomic island of unknown origin containing the antibiotic resistance gene *mecA*[14]. SCC*mec* contains extra genes beyond *mecA*, which include the cytolysin gene *psm-mec*. These may suppress virulence in HA-acquired MRSA strains. At present, twelve unique SCC*mec* types ranging in size from 21–67 kb have been identified. They are designated types I–XII and are distinguished by variation in *mec* and *ccr* gene complexes. SCC*mec* is believed to have originated from close related *Staphylococcus sciuri* species and transferred horizontally to *S. aureus*. Different SCC*mec* genotypes confer different microbiological characteristics, such as different antimicrobial resistance patterns. Different genotypes are also responsible for different types of infections. Types I–III SCC*mec* are large elements that typically contain additional resistance genes and are characteristically isolated from HA-MRSA strains. Conversely, CA-MRSA is associated with types IV and V, which are smaller and lack resistance genes other than *mecA*. Carriage of large plasmids, such as SCC*mec*I–III, is costly to the bacteria, resulting in a compensatory decrease in virulence expression[1]. Entry of *SCCmec* in methicillin-sensitive *S. aureus* (MSSA) produce a number of genetically different MRSA lineages. These genetic variations amongst different MRSA strains probably explain the variability in its virulence and associated MRSA infections[15].

Antibiotic Resistance Patterns and Treatment of MRSA Infection

MRSA strains have been observed to be multi-drug resistant, such as aminoglycosides, macrolides, lincosamides, streptogramins, tetracyclines, *etc.*, which are often used in the treatment of human as well as in veterinary practices. In livestock Beta Lactam antibiotics are frequently used in intramammary infusion therapy. The delay treatment of MRSA infection is must be avoided. The route of administration of an antibiotic varies depending upon the location of the infections. Antibiotics effective against MRSA can be given by IV, oral, or a combination of both. The use of simultaneous treatment with vancomycin or other beta-lactam antibiotic may have a synergistic result. The CA-MRSA as well as HA-MRSA are resistant to traditional anti-staphylococcal beta-lactam antibiotics, such as cephalexin. CA-MRSA has a greater spectrum of antimicrobial susceptibility to sulfa drugs (like co-trimoxazole (trimethoprim/ sulfamethoxazole), tetracyclines (like doxycycline and minocycline) and clindamycin. Linezolid belongs to the newer oxazolidinone class of antibiotics which is effective against both CA-MRSA and HA-MRSA. The Infectious Disease Society of America recommends vancomycin, linezolid, or clindamycin (if susceptible) for treating those with MRSA pneumonia in human being. A fifth-generation cephalosporin, ceftaroline is the beta-lactam antibiotic used to treat MRSA infections in skin and soft tissue or community-acquired pneumonia. Vancomycin and teicoplanin are glycopeptide antibiotics used to treat MRSA infections. Due to low oral absorption of vancomycin and teicoplanin, these should be administered intravenously to control systemic infections. Vancomycin was the drug of choice for treating invasive methicillin-resistant *Staphylococcus aureus* infections during early days. Various newly identified strains of MRSA show antibiotic resistance even to vancomycin and teicoplanin. The first reported strain with complete (>16 μg/ml) resistance to vancomycin, termed vancomycin-resistant *S. aureus* (VRSA) appeared in the United States in 2002[1]. Ceftarolineand ceftobiprole are the first β-lactams with anti-MRSA activity. Use of the combination of daptomycin with an anti-staphylococcal βlactam for refractory MRSA infections has been increasing[8].

There are various studies on MRSA also reported in veterinary field. Bhanderi (2007) carried out antibiotic sensitivity pattern in *Staphylococcus aureus* isolated from mastitic milk samples. They found oxacillin (100 per cent), vancomycin (100 per cent), methicillin (97.67 per cent), cephalothin (93.02 per cent), cephalexin (93.02 per cent) and ciprofloxacin (88.57 per cent) were most effective drugs while streptomycin (39.53 per cent), amoxicillin (38.23 per cent), gentamicin (32.55 per cent), enrofloxacin(27.90 per cent), ampicillin/penicillin (25.58 per cent), oxytetracyclin (20.94 per cent), amikacin (16.28 per cent) and erythromycin (11.62 per cent) were most resistant drugs[16].Haran *et al.* (2011) carried out antimicrobial susceptibility in which two isolates (1.3 per cent) from 150 milk samples were confirmed as methicillin resistant and of the 93 methicillin susceptible isolates 29 per cent were resistant to cloxacillin, 17 per cent to ampicillin, 16 per cent to penicillin and tetracycline each[17].Kumar *et al.* (2011) in his study on 107 isolates of *S. aureus* found fourteen isolates methicillin-resistant, while the remaining (93) were methicillin-susceptible. In his study all isolates (100.0 per cent) were also susceptible to vancomycin[10]. Nimavat (2015) isolated 10 methicillin resistant isolates from bovine mastitis. Out of 10 isolates 2 were methicillin resistant coagulase positive MRSA whereas rests 8 were methicillin resistant coagulase negative staphylococci (MRCONS). In his study he found susceptibility to chloramphenicol (87.09 per cent), oxacillin (83.87

per cent), oxytetracycline (79.03 per cent), methicillin (83.87 per cent), ofloxacin (69.35 per cent) and rifampicin (67.74 per cent)[18]. Ghodasara(2018) evaluated 202 animals and 100 human nasal swabs and isolated 86 (42.57 per cent) and 62 (62 per cent) *Staphylococcus* spp. respectively. He also observed that 8 and 18 isolates were Methicillin-Resistant Coagulase Negative Staphylococci (MRCoNS) while 1 and 2 isolates were Methicillin-Resistant Coagulase Positive Staphylococci (MRCoPS) from animal and humans, respectively. He identified *mecA* gene in 12.03 per cent from animals and 39.21 per cent from humans isolates. He also reported CA-MRS (63.04 per cent) was higher as compare to HA-MRS (28.26 per cent) among both the species[19]. Javia (2018) in his study on bovine mastitis observed 38.72 per cent prevalence of Staphylococcus spp. He also reported that 17.88 per cent isolates were resistant to methicillin[20].

MRSA will mostly always coexist with human and animals. Though some report suggest reduction in the incidence of HA- MRSA in various part of globe, the scientific community still require to do lots future work pertaining to multidrug resistant MRSA. WHO is also recommending the same for the future development of newer drugs and preventive measures.

References

1. Cited from https: //en.wikipedia.org/wiki/Methicillin-resistant *Staphylococcus_aureus* dated: 22.07.2021

2. Green, M. and Bradley, A. 2004. Clinical Forum- Staphylococcus aureus mastitis in cattle. UK Vet. 9: 1-9.

3. Moon, J. S.; Lee, A. R.; Kang, H. M.; Lee, E. S.; Joo, Y. S.; Park, Y. H.; Kim, M. N. and Koo, H. C. 2007. Antibiogram and coagulase diversity in staphylococcal enterotoxin-producing *Staphylococcus aureus* from bovine mastitis. *J. Dairy Sci.* **90:** 1716-1724.

4. Kumar, R.; Yadav, B. R. and Singh, R. S. 2010. Genetic Determinants of antibiotic resistance in *Staphylococcus aureus* isolates from milk of mastitic crossbred cattle. *Curr. Microbiol.***60:** 379–386

5. Wang, Y.; Wu, C. M.; Lu, L. M.; Ren, G. W.; Cao X. Y. and Shen, J. Z. 2008. Macrolide lincosamide- resistant phenotypes and genotypes of *Staphylococcus aureus* isolated from bovine clinical mastitis. *Vet. Microbiol.* **130(1-2):** 118-125.

6. Devriese, L. A. and Hommez, J. 1975. Epidemiology of methicillin-resistant *Staphylococcus aureus* in dairy herds. *Res. Vet. Sci.***19:** 23–27.

7. Lee, J.H. 2003. Methicillin (oxacillin)-resistant *Staphylococcus aureus* strains isolated from major food animals and their potential transmission to humans. *Appl.Environ. Microbiol.* **69(11):** 6489-6494.

8. Lee, A., de Lencastre, H., Garau, J. *et al.*, 2018. Methicillin-resistant *Staphylococcus aureus Nat Rev Dis Primers* **4,** 18033. https: / /doi.org/10.1038/ nrdp.2018.33

9. Vanderhaeghen, W.; Cerpentier, T.; Adriaensen, C.; Vicca, J.; Hermans, K. and Butaye, P. 2010. Methicillin-resistant *Staphylococcus aureus* (MRSA) ST398 associated with chlinical and subclinical mastitis in Belgian cows. *Vet.Microbiol.* **144(1-2):** 166-171.

10. Kumar, R.; Yadav, B. R. and Singh, R. S. 2011. Antibiotic resistance and pathogenicity factors in *Staphylococcus aureus* isolated from mastitic Sahiwal cattle. *J. Biosci.***36:** 175–188.

11. Noskin, G. A.; Rubin, R. J. and Schentag, J. J. 2005. The burden of *Staphylococcus aureus* infections on hospitals in the United States: an analysis of the 2000 and 2001, Nationwide Inpatient Sample Database. *Arch. Intern. Med.***165:** 1756 –1761.

12. Australian Group on Antimicrobial Resistance (AGAR), 2018. AGAR Group- http: / /www.agargroup.org.au.

13. Mitchell, B. G., Collignon, P. J., McCann, R., Wilkinson, I. J. and Wells. 2014. A. A major reduction in hospital-onset *Staphylococcus aureus*bacteremia in Australia12 years of progress: an observational study. *Clin. Infect. Dis.* 59, 969–975.

14. Williamson, D. A., Coombs, G. W. and Nimmo, G. R. 2014 *Staphylococcus aureus* 'Down Under': contemporary epidemiology of *S. aureus* in Australia, New Zealand, and the South West Pacific. *Clin. Microbiol. Infect.* 20, 597–604.

15. Weidenmaier, C. *et al.*, 2008. Differential roles of sortase-anchored surface proteins and wall teichoic acid in *Staphylococcus aureus* nasal colonization. *Int. J. Med. Microbiol.* 298, 505–513.

16. Bhanderi, B. 2007. Isolation, identification, biochemical characterization antibiogram pattern and molecular characterization of *Staphylococcus aureus* from clinical and sub-clinical mastitic milk. M.V.Sc. Thesis, Submitted to Anand Agricultural University, Anand.

17. Haran, K. P.; Godden, S. M.; Boxrud, D.; Jawahir, S.; Bender, J. B. and Sreevatsan, S. 2011. Prevalence and Characterization of *Staphylococcus aureus*, Including Methicillin-Resistant *Staphylococcus aureus*, Isolated from Bulk Tank Milk from Minnesota Dairy Farms.*J. Clin. Microbiol.* **50(3):** 688-695

18. Nimavat, V. R. 2015. Optimization of Multiplex PCR for Identification of MRSA from Bovine Milk Samples and Its Comparison with Conventional Diagnostic Methods. M.V.Sc. Thesis, Submitted to Junagadh Agricultural University, Junagadh.

19. Ghodasara, S. N. 2018. Phenotypic and molecular characterization of staphylococcal cassette chromosome *mec* (*sccmec*) types of methicillin-resistant staphylococci from animal and human origin. Ph.D. Thesis, Submitted to Junagadh Agricultural University, Junagadh.

20. Javia B. B. 2018. Bacteriological studies and molecular detection of major pathogens from subclinical and clinical bovine mastitis. Ph.D. Thesis, Submitted to Junagadh Agricultural University, Junagadh.

Chapter 23

Peste-des-Petits-Ruminants: Current Status and Future Perspectives

D. Muthuchelvan*, Manisha Medhi, R. Mageswary, R. Ashwini Chaple, S. ChandraSekar, and M.A. Ramakrishnan

Indian Veterinary Research Institute, Bengaluru – 560 02, Karnataka
**e-mail: drchelva@gmail.com*

Introduction

Peste-des-petits-ruminants (PPR) is a viral disease of sheep and goats caused by PPR virus (PPRV) with huge economical concern. India harbours 74.26 and 148.88 million of sheep and goats, respectively (20th Livestock census, 2012; http://dahd.nic.in/dahd/WriteReadData/Livestock.pdf). The disease threatens 70 per cent of the landless labourers, small and medium farmers who rears sheep and goats for their livelihood. As the disease affects the poor sections of the society, international agencies such as FAO and OIE set a target for global eradication by 2030 (http://www.fao.org/ppr/en/). The disease causes up to 100 per cent mortality in naive animals and productive loss. It was estimated that PPR alone cause 1,871 million Indian Rupees losses, annually (Singh 2012; Vashist *et al.*, 2021). As one of the forerunners in the rinderpest (RP) research and significant contribution in the global eradication of the disease, consequently, IVRI, Mukteswar developed live attenuated vaccine (Sungri96) and monoclonal antibody based diagnostics (Singh *et al.*, 2004b,c). The above vaccine and diagnostics are supplied throughout India as well as neighbouring countries. Availability of these technologies helped India to initiate mass vaccination campaign in 2004 which brought down the incidence of the disease significantly. Further, Government of India has recently launched a national control program on PPR (NCP-PPR) (http://dahd.nic.in/dahd/WriteReadData/Livestock per cent 20Health.pdf).

The Disease

PPR is a disease of small ruminants which resembles rinderpest of cattle. The incubation period is 3-6 days, followed by high fever, oculonasal discharges, pneumonia, stomatitis, and inflammation of gastrointestinal tract leading to severe diarrhoea followed by death or recovery (Balamurugan *et al.*, 2014b; Sen *et al.*, 2010; Zahur *et al.*, 2008). The disease causes more severe lesions in goats than sheep. Although, the reason for this host specificity is not fully understood, it could be due to difference in genetic makeup and/or receptor distributions of the host. Different levels of SignalingLymphocyte Activation Molecule (SLAM;CD150) mRNA could influence the virus replication in different species (Pawar *et al.*, 2008a). Another study examined the replication of PPRV in peripheral blood mononuclear cells (PBMCs) of Indian goats and water buffalo and demonstrated that the level of TLR3 and TLR7 and downstream signalling molecules correlate with susceptibility(Dhanasekaran *et al.*, 2014)). Incidence in other species such as cattle, pig, camel, buffalo, lion and captive wild small ruminants were reported but are not contributing to the disease epidemiology. Experimental infection of cattle with PPRV revealed that the nucleic acid could be detected up to 397 days post infection (Sen *et al.*, 2014).

No carrier state or persistent infections have been reported so far. However, disease may circulate in subclinical forms in sheep and goats of endemic region. The recovered animals are immune for lifelong. The cytokine expression profile (IL-4 and IFN-γ) of vaccinated and infected goats were found to have a unique biphasic response of IL-4 expression with up-regulation of IFN-γ on 7th days post vaccination (Patel *et al.*, 2012). PPRV was reported to induce apoptosis *in vitro* in goat PBMCs (Mondal *et al.*, 2001).

Although presence of PPR like disease has been suspected earlier in retrospective study (Taylor *et al.*, 2002) its presence was first confirmed in India in1987 from Arasur village of Villupuram district of Tamil Nadu (Shaila *et al.*, 1989). Currently, PPR outbreaks are being reported regularly from different parts of the country (Chauhan *et al.*, 2011; Kerur *et al.*, 2008; Muthuchelvan *et al.*, 2014; Nanda *et al.*; 1996; Raghavendra *et al.*, 2008; Singh *et al.*, 2004a). Sero-prevalence of antibody against PPRV in goats and sheep was 43.56 per cent for the whole country and 11.63 per cent for the North-East states (Balamurugan *et al.*, 2011, 2014a). Antibodies to PPRV in cattle and buffalos were found to be 4.58 per cent in India (Balamurugan *et al.*, 2012a). Molecular epidemiological studies confirm the circulation of lineage IV in India (Balamurugan *et al.*, 2010; Dhar *et al.*, 2002; Muthuchelvan *et al.*, 2014; Shaila *et al.*, 1996). Presence of mixed infection with bluetongue virus (Mondal *et al.*, 2009), goatpox virus (Malik *et al.*, 2011) and Orf virus (Saravanan *et al.*, 2007) have also been reported.

The Virus

Peste-des-petits-ruminants virus (PPRV) is a member of the genus *Morbillivirus.* subfamily *Paramyxovirinae,* family *Paramyxoviridae*in the order *Mononegavirales*(International Committee on Taxonomy of Viruses and King 2012). Since the virus is enveloped one, it is easily inactivated under sunlight and with many chemicals. The genome of PPRV is a single-stranded RNA, with negative polarity with the size of 15948 bp. The genome is organized into six transcriptional units and each encodes at least one non overlapping protein: the nucleocapsid (N), the matrix (M), the polymerase or large (L), the phosphoprotein (P), and two envelope proteins haemagglutin (H) and fusion (F). The P protein uses alternate expression strategies to code for two nonstructural proteins V and C (Muthuchelvan *et al.,* 2006). The PPRV genes arranged from 3' to 5' on the genome is in an order of N-P-M-F-H-L separated by intergenic region which is CTT in most cases. At the 3' and 5' end of the genome there is a leader (52 nucleotides) and trailer (37 nucleotides) region contain promoter functions (Bailey *et al.,* 2005). Two heptad repeats (HR1 and HR2) of the fusion protein was shown to form a six-helix and trimeric coiled-coil bundle for initiation of fusion activity (Rahaman *et al.,* 2003). Host-pathogen interaction study with vaccine virus (Sungri96) and goat PBMCs revealed the dysregulation of immune regulatory pathways (Manjunath *et al.,* 2015). Establishment of an *in vitro* system using Sungri96 virus shown that RNP complex is actively synthesis mRNA (Yunus and Shaila 2012). PPRV multiplication was reduced significantly in SLAM receptors suppressed with siRNA in B95a (Pawar *et al.,* 2008b).Transiently expressed PPRV F glycoprotein induces cell fusion in the absence of H protein (Seth and Shaila 2001). The H protein was found to have potential T cell determinant(s) at amino acids position 123-137 and 242-609 (Sinnathamby *et al.,* 2001).

Molecular Epidemiology

Initially, based on the fusion gene (F), PPRV was classified in to 4 lineages I–IV (Dhar *et al.,* 2002; Shaila *et al.,* 1996). These studies identified lineage I-III viruses across several countries of Africa and lineage IV (Asian lineage) mainly in Middle East and Asia and was supported by other works (Banyard *et al.,* 2010; Dhar *et al.,* 2002; Ozkul *et al.,* 2002). Currently, phylogenetic analysis based on the nucleoprotein gene (N) preferred over F gene analysis due to its better molecular epidemiological pattern (Kwiatek *et al.,* 2007). In the recent past, many reports indentified the presence of Lineage IV in many African countries (Maganga *et al.,* 2013; Salami *et al.,* 2014).So far, only lineage IV viruses have been reported from India. The recent outbreak strain at Tripura had 99.2 to 99.6 per cent nucleotide identities with the Bangladesh strains (Muthuchelvan *et al.,* 2014). This study confirms the transboundary circulation of PPRV between the countries and warrants immediate vaccination throughout the border regions.

Vaccine

Owing to cross-protection between RPV and PPRV, an attenuated tissue culture RP vaccine (Plowright's strain) was used in many countries for the control of PPR in small ruminants. However, the global eradication of RP since May 2011 necessitated the use of only PPRV vaccine. The first homologous vaccine against PPR was developed in 1987 using a Nigerian isolate (Nigeria/75/1). Subsequently, three PPRV vaccines were developed in India using three isolates of lineage IV (PPRV/Sungri/96, PPRV/Arasur/87, and PPRV/Coimbatore/97) (Palaniswami *et al.*, 2005; Singh *et al.*, 2004a). All the three vaccines were found to be equally efficacious in protecting the animals (Santhosh *et al.*, 2013; Saravanan *et al.*, 2010). However, the replication cycle and monoclonal reactivity of Arasur/87 virus is different from Sungri/96 virus (Singh *et al.*, 2010)Currently, vaccine prepared from Sungri/96 is commercially available and is being used in the ongoing control program in India. This vaccine is safe, potent and confers life-long immunity. Further, this vaccine will not induce immunosuppression and is safe for pregnant animal also (Rajak *et al.*, 2005). Cold chain maintenance is the major concern of live attenuated vaccines in tropical conditions. To overcome this problem, attempts were made to develop thermo adapted vaccine and shown to have higher shelf life at high temperature (>40 °C) (Balamurugan *et al.*, 2014c; Riyesh *et al.*, 2011). Being a live attenuated vaccine, animals mount strong antibody response against all the viral proteins at different level and therefore differentiating the infected from the vaccinated animals (DIVA) would be difficult. Efforts are currently underway to develop a DIVA vaccine for PPRV at our lab. Combined vaccines *viz.*, PPR and goatpox(Hosamani *et al.*, 2006), PPR and sheeppox(Chaudhary *et al.*, 2009) have been shown as promising candidates for mass vaccination.Appropriate period of vaccination in kids was found to be four months (Balamurugan *et al.*, 2012b). Attempt was made to develop a subunit vaccine by expressing the H protein of PPRV in peanut plants (*Arachis hypogea*) and found to elicit neutralizing antibody responses in sheep (Khandelwal *et al.*, 2011). Construction of two recombinant viruses using Modified Vaccinia Angara expressing the full length -F and H proteins protected goats against virulent challenge (Chandran *et al.*, 2010). Recent study compared the Sungri/96 vaccine virus with Nigeria/75/1 and found that Sungri/96 vaccine elicited very good humoral and CMI responses and was able to protect all circulating lineages (Hodgson *et al.*, 2018).

Diagnostics of PPR

Laboratory confirmation of the disease is usually done through virus isolation and virus neutralization assay, which are time consuming and laborious. Several assays have been described to detect virus-specific antibodies or viral antigen.

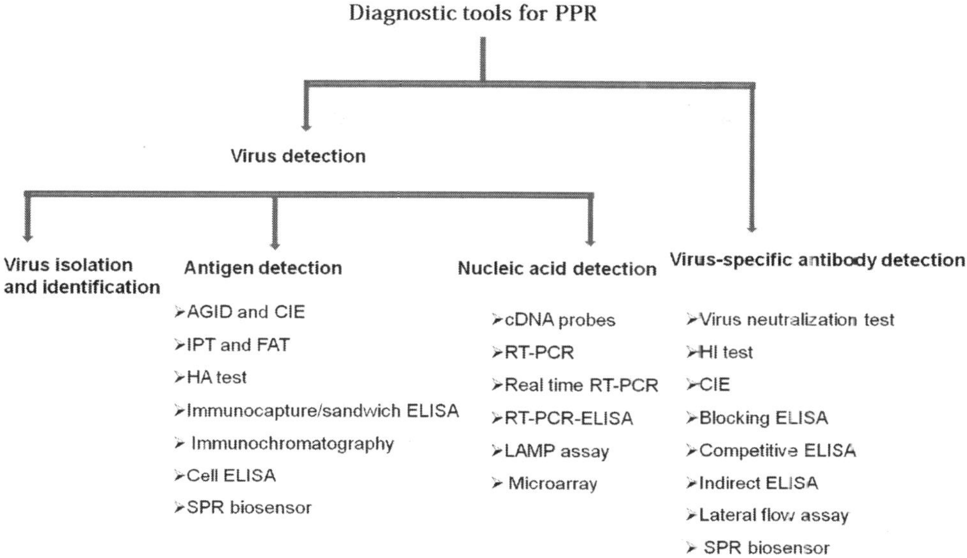

Various diagnostic tools used in PPR research Source: (Santhamani *et al.*, 2016)

A) Detection of PPRV Antibody

A monoclonal antibody based competitive ELISA (cELISA) kit for detection of antibodies against PPRV has been developed in our laboratotory(Singh *et al.*, 2004a). The assay is suitable for both vaccine seromonitoring and monitoring of post infection. The test was compared with VNT and found to be 92.2 per cent and 98.84 per cent, of specificity and sensitivity, respectively.

A polyclonal antibody based indirect ELISA was developed for detection of antibodies against PPRV in the serum samples (Balamurugan *et al.*, 2007). The performance of the assay was comparable with that of c-ELISAs.

B) Detection of PPRV Antigen

Virus Isolation

Virus isolation remains the *"gold standard"* for the diagnosis of PPR. Blood collected at the height of the fever is the best material for this purpose. The nasal or ocular swabs or 10 per cent tissue suspension can also be used for virus isolation. The PPR virus can be propagated *in vitro* in primary bovine and sheep cells, as well as established cell lines like Vero (African green monkey kidney cells) and B95a (Marmoset B-lymphoblastoid cells) (Sreenivasa *et al.*, 2006). In most of these cells, the PPR virus manifests morbillivirus-specific CPE by day 3-5. In some cases up to five blind passages are needed.

Sandwich ELISA

The sandwich-ELISA for the detection of PPR antigen in the clinical samples was developed at IVRI, Mukteswar. This kit uses a monoclonal antibody directed against an N protein of PPR virus. The test is compared with internationally accepted IC-ELISA (BDSL) and found to have sensitivity and specificity of 89 per cent and 93 per cent, respectively (Singh *et al.*, 2004b). The recombinant N protein based ELISA was also evaluated (Yadav *et al.*, 2009). Recently, IVRI-Mukteswar lab has released a baculovirus expressed recombinant N protein based sandwich ELISA kit.

C) Detection of PPRV Nucleic Acid

Reverse Transcription-Polymerase Chain Reaction (RT-PCR) is the method of choice for detecting PPRV genome in clinical samples. RT-PCRs have been reported for detection and differential diagnosis of RP and PPR viruses (Balamurugan *et al.*, 2006; Couacy-Hymann *et al.*, 2002; Forsyth and Barrett 1995). Diagnostic PCR based on other genes are available (Brindha *et al.*, 2001; George *et al.*, 2006) but are no in common use.

N-Gene based PCR-ELISA has developed at IVRI, Mukteswar to detect PPRV from clinical samples especially from critical samples. The test can detect viral RNA in infected tissue culture fluid with a titre as low as 0.01 $TCID_{50}$ (Saravanan *et al.*, 2004).

Real-time RT-PCR

Real time RT- PCR recently been used for quantification and diagnosis of PPR virus from clinical samples. M gene-based hydrolysis probe (TaqMan), one step and two-step SYBR Green I based real-time RT-PCR assays targeting the M gene of PPRV are in use (Abera and Thangavelu 2014; Balamurugan *et al.*, 2012c). All the assays were sensitive and could be applied in clinical diagnosis. Recently, a novel non-amplification technique was developed to detect nucleic acids of PPRV in which two probes complementary to the target sequences - one conjugated to magnetic microparticles and the second to gold nanoparticles labelled with horseradish peroxidased(Tao *et al.*, 2013). The assay has great potential due to quick performance (45 min) and not requiring expensive instrumentations *etc.*

Lateral Flow or Dipstick Test

Lateral flow test also known as chromatographic strip test on the principle of lateral flow of antigen-antibody complex resulting in colour development while reacting with antibody conjugated to gold. Similar assay has currently developed in India and validation process is under the progress.

Loop Mediated Isothermal Amplification (LAMP) Assay

LAMP assay for rapid detection of PPRV from clinical samples has recently been standardized at IVRI, Mukteswar. The assay is based on targeting conserved region of 'N' gene of PPRV. Hydroxy naphthaol blue – a DNA binding dye was used for visual detection of the positive samples. The original colour of dye is violet but in positive reaction, it turns to sky blue due to pH change of reaction whereas the colour remains unchanged in negative reaction. The primer of LAMP is specific for PPR and not cross reactive with other related viruses of sheep and goats. The assay is 100 and 1000 times more sensitive than RT-PCR and s-ELISA, respectively (Dadas *et al.*, 2012).

Control Program in India

For effective PPR control, focused vaccinations in high-risk populations and mass vaccination are necessary to achieve 70 to 80 per cent of herd immunity. The control strategies should follow the "bottom-up" approach (farmers to field veterinarians to policy makers) (Singh 2011). Besides, strengthening of PPR vaccine production and quality control units, disease diagnostic laboratories, infrastructure for field Veterinarians are also key factors for effective control strategies. The NCP-PPR control programme started in 2010 and operates in different state/province: Kerala, Tamil Nadu, Karnataka, Andhra Pradesh, Maharashtra, Goa and Union Territories of Daman and Diu, Nagar Haveli, Andaman and Nicobar Islands and Puducherry. The aim of the programme is to do intense vaccination all susceptible goats and sheep and their three subsequent generations. The impact of the programme is already realized with significant reduction of PPR outbreaks in states like Karnataka and Andhra Pradesh.

References

1. Abera T, Thangavelu A. 2014. Development of a two-step SYBR Green I based real time RT-PCR assay for detecting and quantifying peste des petits ruminants virus in clinical samples. *J. Virol. Methods*. 209: 25–29
2. Bailey D, Banyard A, Dash P, Ozkul A, Barrett T. 2005. Full genome sequence of peste des petits ruminants virus, a member of the Morbillivirus genus. *Virus Res.* 110(1–2): 119–24
3. Balamurugan V, Das S, Raju DSN, Chakravarty I, Nagalingam M, *et al.*, 2014a. Prevalence of peste des petits ruminants in goats in North-East India. *Virusdisease*. 25(4): 488–92
4. Balamurugan V, Hemadri D, Gajendragad MR, Singh RK, Rahman H. 2014b. Diagnosis and control of peste des petits ruminants: a comprehensive review. *Virusdisease*. 25(1): 39–56
5. Balamurugan V, Krishnamoorthy P, Veeregowda BM, Sen A, Rajak KK, *et al.*, 2012a. Seroprevalence of Peste des petits ruminants in cattle and buffaloes from Southern Peninsular India. *Trop. Anim. Health Prod.* 44(2): 301–6
6. Balamurugan V, Saravanan P, Sen A, Rajak KK, Bhanuprakash V, *et al.*, 2011. Sero-epidemiological study of peste des petits ruminants in sheep

and goats in India between 2003 and 2009. *Rev. Sci. Tech. Int. Off. Epizoot.* 30(3): 889–96

7. Balamurugan V, Sen A, Saravanan P, Singh RP, Singh RK, *et al.*, 2006. One-step multiplex RT-PCR assay for the detection of peste des petits ruminants virus in clinical samples. *Vet. Res. Commun.* 30(6): 655–66
8. Balamurugan V, Sen A, Venkatesan G, Bhanuprakash V, Singh RK. 2014c. Protective immune response of live attenuated thermo-adapted peste des petits ruminants vaccine in goats. *Virusdisease.* 25(3): 350–57
9. Balamurugan V, Sen A, Venkatesan G, Rajak KK, Bhanuprakash V, Singh RK. 2012b. Study on passive immunity: Time of vaccination in kids born to goats vaccinated against Peste des petits ruminants. *Virol. Sin.* 27(4): 228–33
10. Balamurugan V, Sen A, Venkatesan G, Yadav V, Bhanot V, *et al.*, 2012c. A rapid and sensitive one step-SYBR green based semi quantitative real time RT-PCR for the detection of peste des petits ruminants virus in the clinical samples. *Virol. Sin.* 27(1): 1–9
11. Balamurugan V, Sen A, Venkatesan G, Yadav V, Bhanuprakash V, Singh RK. 2010. Isolation and identification of virulent peste des petits ruminants viruses from PPR outbreaks in India. *Trop. Anim. Health Prod.* 42(6): 1043–46
12. Balamurugan V, Singh RP, Saravanan P, Sen A, Sarkar J, *et al.*, 2007. Development of an indirect ELISA for the detection of antibodies against Peste-des-petits-ruminants virus in small ruminants. *Vet. Res. Commun.* 31(3): 355–64
13. Banyard AC, Parida S, Batten C, Oura C, Kwiatek O, Libeau G. 2010. Global distribution of peste des petits ruminants virus and prospects for improved diagnosis and control. *J. Gen. Virol.* 91(Pt 12): 2885–97
14. Brindha K, Raj GD, Ganesan PI, Thiagarajan V, Nainar AM, Nachimuthu K. 2001. Comparison of virus isolation and polymerase chain reaction for diagnosis of peste des petits ruminants. *Acta Virol.* 45(3): 169–72
15. Chandran D, Reddy KB, Vijayan SP, Sugumar P, Rani GS, *et al.*, 2010. MVA recombinants expressing the fusion and hemagglutinin genes of PPRV protects goats against virulent challenge. *Indian J. Microbiol.* 50(3): 266–74
16. Chaudhary SS, Pandey KD, Singh RP, Verma PC, Gupta PK. 2009. A vero cell derived combined vaccine against sheep pox and Peste des Petits ruminants for sheep. *Vaccine.* 27(19): 2548–53
17. Chauhan HC, Lambade PS, Sen A, Dadawala AI, Ranaware PB, *et al.*, 2011. The use of pathological and histopathological techniques in the diagnosis of peste des petits ruminants in India. *Vet. Ital.* 47(1): 41–47
18. Couacy-Hymann E, Roger F, Hurard C, Guillou JP, Libeau G, Diallo A. 2002. Rapid and sensitive detection of peste des petits ruminants virus by a polymerase chain reaction assay. *J. Virol. Methods.* 100(1–2): 17–25

19. Dadas RC, Muthuchelvan D, Pandey AB, Rajak KK, Sudhakar SB, *et al.*, 2012. Development of loop-mediated isothermal amplification (LAMP) assay for rapid detection of peste des petits ruminants virus (PPRV) genome from clinical samples. *Indian J. Comp. Microbiol. Immunol. Infect. Dis.* 33(1–2): 7–13
20. Dhanasekaran S, Biswas M, Vignesh AR, Ramya R, Raj GD, *et al.*, 2014. Toll-like receptor responses to Peste des petits ruminants virus in goats and water buffalo. *PloS One*. 9(11): e111609
21. Dhar P, Sreenivasa BP, Barrett T, Corteyn M, Singh RP, Bandyopadhyay SK. 2002. Recent epidemiology of peste des petits ruminants virus (PPRV). *Vet. Microbiol.* 88(2): 153–59
22. Forsyth MA, Barrett T. 1995. Evaluation of polymerase chain reaction for the detection and characterisation of rinderpest and peste des petits ruminants viruses for epidemiological studies. *Virus Res.* 39(2–3): 151–63
23. George A, Dhar P, Sreenivasa BP, Singh RP, Bandyopadhyay SK. 2006. The M and N genes-based simplex and multiplex PCRs are better than the F or H gene-based simplex PCR for Peste-des-petits-ruminants virus. *Acta Virol.* 50(4): 217–22
24. Hodgson S, Moffat K, Hill H, Flannery JT, Graham SP, *et al.*, 2018. Comparison of the Immunogenicities and Cross-Lineage Efficacies of Live Attenuated Peste des Petits Ruminants Virus Vaccines PPRV/Nigeria/75/1 and PPRV/Sungri/96. *J. Virol.* 92(24): e01471-18
25. Hosamani M, Singh SK, Mondal B, Sen A, Bhanuprakash V, *et al.*, 2006. A bivalent vaccine against goat pox and Peste des Petits ruminants induces protective immune response in goats. *Vaccine.* 24(35–36): 6058–64
26. International Committee on Taxonomy of Viruses, King AMQ, eds. 2012. *Virus Taxonomy: Classification and Nomenclature of Viruses: Ninth Report of the International Committee on Taxonomy of Viruses.* London/; Waltham, MA: Academic Press
27. Kerur N, Jhala MK, Joshi CG. 2008. Genetic characterization of Indian peste des petits ruminants virus (PPRV) by sequencing and phylogenetic analysis of fusion protein and nucleoprotein gene segments. *Res. Vet. Sci.* 85(1): 176–83
28. Khandelwal A, Renukaradhya GJ, Rajasekhar M, Sita GL, Shaila MS. 2011. Immune responses to hemagglutinin-neuraminidase protein of peste des petits ruminants virus expressed in transgenic peanut plants in sheep. *Vet. Immunol. Immunopathol.* 140(3–4): 291–96
29. Kwiatek O, Minet C, Grillet C, Hurard C, Carlsson E, *et al.*, 2007. Peste des petits ruminants (PPR) outbreak in Tajikistan. *J. Comp. Pathol.* 136(2–3): 111–19

30. Maganga GD, Verrier D, Zerbinati RM, Drosten C, Drexler JF, Leroy EM. 2013. Molecular typing of PPRV strains detected during an outbreak in sheep and goats in south-eastern Gabon in 2011. *Virol. J.* 10: 82

31. Malik YS, Singh D, Chandrashekar KM, Shukla S, Sharma K, *et al.*, 2011. Occurrence of dual infection of peste-des-petits-ruminants and goatpox in indigenous goats of central India. *Transbound. Emerg. Dis.* 58(3): 268–73

32. Manjunath S, Kumar G, Mishra B, Mishra B, Sahoo A, *et al.*, 2015. Genomic analysis of host - Peste des petits ruminants vaccine viral transcriptome uncovers transcription factors modulating immune regulatory pathways. *Vet. Res.* 46(1): 15

33. Mondal B, Sen A, Chand K, Biswas SK, De A, *et al.*, 2009. Evidence of mixed infection of peste des petits ruminants virus and bluetongue virus in a flock of goats as confirmed by detection of antigen, antibody and nucleic acid of both the viruses. *Trop. Anim. Health Prod.* 41(8): 1661–67

34. Mondal B, Sreenivasa BP, Dhar P, Singh RP, Bandyopadhyay SK. 2001. Apoptosis induced by peste des petits ruminants virus in goat peripheral blood mononuclear cells. *Virus Res.* 73(2): 113–19

35. Muthuchelvan D, De A, Debnath B, Choudhary D, Venkatesan G, *et al.*, 2014. Molecular characterization of peste-des-petits ruminants virus (PPRV) isolated from an outbreak in the Indo-Bangladesh border of Tripura state of North-East India. *Vet. Microbiol.* 174(3–4): 591–95

36. Muthuchelvan D, Sanyal A, Sarkar J, Sreenivasa BP, Bandyopadhyay SK. 2006. Comparative nucleotide sequence analysis of the phosphoprotein gene of peste des petits ruminants vaccine virus of Indian origin. *Res. Vet. Sci.* 81(1): 158–64

37. Nanda YP, Chatterjee A, Purohit AK, Diallo A, Innui K, *et al.*, 1996. The isolation of peste des petits ruminants virus from northern India. *Vet. Microbiol.* 51(3–4): 207–16

38. Ozkul A, Akca Y, Alkan F, Barrett T, Karaoglu T, *et al.*, 2002. Prevalence, distribution, and host range of Peste des petits ruminants virus, Turkey. *Emerg. Infect. Dis.* 8(7): 708–12

39. Palaniswami KS, Thangavelu A, Velmurugan R. 2005. Development of Thermostable Peste Des Petits Ruminants (PPR) Virus Vaccine and Assessment of Molecular Changes in the F Gene. In *Applications of Gene-Based Technologies for Improving Animal Production and Health in Developing Countries*, eds. HPS Makkar, GJ Viljoen, pp. 673–78. Berlin/Heidelberg: Springer-Verlag

40. Patel A, Rajak KK, Balamurugan V, Sen A, Sudhakar SB, *et al.*, 2012. Cytokines expression profile and kinetics of Peste des petits ruminants virus antigen and antibody in infected and vaccinated goats. *Virol. Sin.* 27(4): 265–71

41. Pawar RM, Dhinakar Raj G, Balachandran C. 2008a. Relationship between the level of signaling lymphocyte activation molecule mRNA and replication of Peste-des-petits-ruminants virus in peripheral blood mononuclear cells of host animals. *Acta Virol.* 52(4): 231–36

42. Pawar RM, Raj GD, Kumar TMAS, Raja A, Balachandran C. 2008b. Effect of siRNA mediated suppression of signaling lymphocyte activation molecule on replication of peste des petits ruminants virus in vitro. *Virus Res.* 136(1–2): 118–23

43. Raghavendra AG, Gajendragad MR, Sengupta PP, Patil SS, Tiwari CB, *et al.*, 2008. Seroepidemiology of peste des petits ruminants in sheep and goats of southern peninsular India. *Rev. Sci. Tech. Int. Off. Epizoot.* 27(3): 861–67

44. Rahaman A, Srinivasan N, Shamala N, Shaila MS. 2003. The fusion core complex of the peste des petits ruminants virus is a six-helix bundle assembly. *Biochemistry.* 42(4): 922–31

45. Rajak KK, Sreenivasa BP, Hosamani M, Singh RP, Singh SK, *et al.*, 2005. Experimental studies on immunosuppressive effects of peste des petits ruminants (PPR) virus in goats. *Comp. Immunol. Microbiol. Infect. Dis.* 28(4): 287–96

46. Riyesh T, Balamurugan V, Sen A, Bhanuprakash V, Venkatesan G, *et al.*, 2011. Evaluation of efficacy of stabilizers on the thermostability of live attenuated thermo-adapted Peste des petits ruminants vaccines. *Virol. Sin.* 26(5): 324–37

47. Salami H, Croville G, Kwiatek O, Mariette J, Klopp C, *et al.*, 2014. Complete Genome Sequence of a Field Strain of Peste des Petits Ruminants Virus Isolated during 2010-2014 Epidemics in Senegal. *Genome Announc.* 2(5):

48. Santhamani R, Singh RP, Njeumi F. 2016. Peste des petits ruminants diagnosis and diagnostic tools at a glance: perspectives on global control and eradication. *Arch. Virol.* 161(11): 2953–67

49. Santhosh AK, Gomes AR, Hegde R, Rathnamma D, Veeregowda BM, *et al.*, 2013. Comparative immunogenicity of two peste des petitis ruminants (PPR) vaccines in South Indian sheep and goats under field conditions. *Indian J. Virol. Off. Organ Indian Virol. Soc.* 24(3): 373–79

50. Saravanan P, Balamurugan V, Sen A, Sarkar J, Sahay B, *et al.*, 2007. Mixed infection of peste des petits ruminants and orf on a goat farm in Shahjahanpur, India. *Vet. Rec.* 160(12): 410–12

51. Saravanan P, Sen A, Balamurugan V, Rajak KK, Bhanuprakash V, *et al.*, 2010. Comparative efficacy of peste des petits ruminants (PPR) vaccines. *Biol. J. Int. Assoc. Biol. Stand.* 38(4): 479–85

52. Saravanan P, Singh RP, Balamurugan V, Dhar P, Sreenivasa BP, *et al.*, 2004. Development of a N gene-based PCR-ELISA for detection of Peste-des-petits-ruminants virus in clinical samples. *Acta Virol.* 48(4): 249–55

53. Sen A, Saravanan P, Balamurugan V, Bhanuprakash V, Venkatesan G, *et al.*, 2014. Detection of subclinical peste des petits ruminants virus infection in experimental cattle. *Virusdisease*. 25(3): 408–11

54. Sen A, Saravanan P, Balamurugan V, Rajak KK, Sudhakar SB, *et al.*, 2010. Vaccines against peste des petits ruminants virus. *Expert Rev. Vaccines*. 9(7): 785–96

55. Seth S, Shaila MS. 2001. The hemagglutinin-neuraminidase protein of peste des petits ruminants virus is biologically active when transiently expressed in mammalian cells. *Virus Res.* 75(2): 169–77

56. Shaila MS, Purushothaman V, Bhavasar D, Venugopal K, Venkatesan RA. 1989. Peste des petits ruminants of sheep in India. *Vet. Rec.* 125(24): 602

57. Shaila MS, Shamaki D, Forsyth MA, Diallo A, Goatley L, *et al.*, 1996. Geographic distribution and epidemiology of peste des petits ruminants virus. *Virus Res.* 43(2): 149–53

58. Singh RP. 2011. Control strategies for peste des petits ruminants in small ruminants of India. *Rev. Sci. Tech.-OIE*. 30(3): 879

59. Singh RP. 2012. Strategic control of peste des petits ruminants. *Vet. Livest. Sect. Bluepr. Capacity Build.* 327–45

60. Singh RP, De UK, Pandey KD. 2010. Virological and antigenic characterization of two Peste des Petits Ruminants (PPR) vaccine viruses of Indian origin. *Comp. Immunol. Microbiol. Infect. Dis.* 33(4): 343–53

61. Singh RP, Saravanan P, Sreenivasa BP, Singh RK, Bandyopadhyay SK. 2004a. Prevalence and distribution of peste des petits ruminants virus infection in small ruminants in India. *Rev. Sci. Tech. Int. Off. Epizoot.* 23(3): 807–19

62. Singh RP, Sreenivasa BP, Dhar P, Bandyopadhyay SK. 2004b. A sandwich-ELISA for the diagnosis of Peste des petits ruminants (PPR) infection in small ruminants using anti-nucleocapsid protein monoclonal antibody. *Arch. Virol.* 149(11): 2155–70

63. Singh RP, Sreenivasa BP, Dhar P, Shah LC, Bandyopadhyay SK. 2004c. Development of a monoclonal antibody based competitive-ELISA for detection and titration of antibodies to peste des petits ruminants (PPR) virus. *Vet. Microbiol.* 98(1): 3–15

64. Sinnathamby G, Renukaradhya GJ, Rajasekhar M, Nayak R, Shaila MS. 2001. Immune responses in goats to recombinant hemagglutinin-neuraminidase glycoprotein of Peste des petits ruminants virus: identification of a T cell determinant. *Vaccine*. 19(32): 4816–23

65. Sreenivasa BP, Singh RP, Mondal B, Dhar P, Bandyopadhyay SK. 2006. Marmoset B95a cells: a sensitive system for cultivation of Peste des petits ruminants (PPR) virus. *Vet. Res. Commun.* 30(1): 103–8

66. Tao C, Li G, Wang Y, Huang H. 2013. Enzymatic reporting of peste des petits ruminants virus genes ligating two specific probes on nanoparticles. *Biotechnol. Lett.* 35(4): 613–18

67. Taylor WP, Diallo A, Gopalakrishna S, Sreeramalu P, Wilsmore AJ, *et al.*, 2002. Peste des petits ruminants has been widely present in southern India since, if not before, the late 1980s. *Prev. Vet. Med.* 52(3–4): 305–12

68. Vashist VS, Yadav AK, Rajak KK, Sekar SC, Ramakrishnan MA, *et al.*, 2021. Flock Level Economic Loss Due to Peste Des Petits Ruminants Outbreak in Transhumance Sheep and Goat Population in Himachal Pradesh (India). *Indian J. Anim. Res.* (Of):

69. Yadav V, Balamurugan V, Bhanuprakash V, Sen A, Bhanot V, *et al.*, 2009. Expression of Peste des petits ruminants virus nucleocapsid protein in prokaryotic system and its potential use as a diagnostic antigen or immunogen. *J. Virol. Methods.* 162(1–2): 56–63

70. Yunus M, Shaila MS. 2012. Establishment of an in vitro transcription system for Peste des petits ruminant virus. *Virol. J.* 9: 302

71. Zahur AB, Irshad H, Hussain M, Ullah A, Jahangir M, *et al.*, 2008. The epidemiology of peste des petits ruminants in Pakistan. *Rev. Sci. Tech. Int. Off. Epizoot.* 27(3): 877–84

Chapter 24

Applications of Nanobodies in Diagnosis and Control of Infectious Diseases

Ajit Singh

Department of Veterinary Microbiology, Lala Lajpat Rai University of Veterinary and Animal Sciences, Hisar – 125 004, Haryana
e-mail: adjitprinik@msn.com

General Introduction

The advent of 21st century heralded the increase in the cataclysmic climate change events, including the enhanced pace of emergence, and spread of new infectious agents, and re-emergence and geospatial re-distribution of altered forms of the pre-existing agents. Anthropogenic activities have fuelled these changes, and therefore only human intervention can mitigate or slow down the disastrous consequences. The current pandemic caused by SARS-CoV-2 is a whistle-blower to alert and awaken people world-wide. Since it is assumed that more of such events are simmering in Nature, wisdom lies in our preparedness for facing future challenges effectively. This would require developing, refining and timely deployment of the early warning systems, diagnostics, preventive and control measures, in addition to inter-sectoral coordination and cooperation at local, national, regional and global levels.

World Animal Health Organization (OIE) has listed 117 animal diseases, infections and infestations listed for the year 2020 [1]. Most of the infectious diseases of animals are not only of veterinary, but also of medical and public health concern, because the agents can infect multiple host species, often crossing or jumping the species barriers. Technology promises and provides solutions to problems. The present-day technology platforms and innovations have provided us with vaccines, diagnostics and other tools to understand, slow down and control the current pandemic in much shorter time and more effectively than the previous pandemics in the history of mankind. However, the efforts need to be ceaseless to face any future challenges. Unfortunately, due to several reasons, including less availability of funds, technological inputs for advancement in animal health sector has continued to lag behind those of human health sector. Paucity of species-specific research reagents, immunobiologicals, molecular and other investigational materials has chronically hampered the progress in the animal health sciences until recently. During past decade, advocacy for 'One Health' has gained momentum, and research priorities are being recast in that frame, which isgradually creating a better and balanced research ecosystem to address the threats of emerging infectious diseases in future.

The engineered antibodies alone as biopharmaceuticals are accounting for more than 30 per cent of revenues in the biotech market at present. The demand for veterinary biologicals, particularly those employed in diagnostics and vaccines has also shown increasing trends for past two decades. The small-sized recombinant antibody formats represent one such class of biological products that have beensuggested as viable alternatives to antisera and monoclonal antibodies for diagnostics, therapeutics and research purposes. Single-domain antibodies (dAbs), also called 'nanobodies' (Nbs) are the smallest of such constructs, often produced as libraries of 10^6-10^{12} size by phage display technology or other display platforms [2-4]. The present discussion is about the Nb- based constructs that represent a class of tools for improved diagnostics, vaccine designs, therapeutics, and research advancement for faster and effective control of infectious diseases of human and animals.

Brief Description of Nanobodies: Production and Features

Nanobodies® ('the nano-sized antibodies'; abbreviated as Nbs;proprietary name coined by Ablynx®; also known as single-domain or simply domain antibodies (dAbs)); are among the smallest antigen (Ag)-binder molecules (12-17 kDa in size), derived and designed from only VH-domain encoding gene segments (called VHH gene segment) of the 'heavy-chain antibody' molecules (HCAbs) of camelids or sharks, and produced by a set of rDNA techniques as a vast collection (known as a library) in a prokaryotic or a eukaryotic host expression system. Phage display and yeast display libraries of Nbs are among the most popular platforms. Immune, naïve and synthetic libraries have been produced as various display platforms since the first phage display library of Nbs produced in 1998 [5]. Many new proprietary technologies for novel Nb-based constructs have recently been announced and reviewed [6]. Ag-specific Nbs are selected from the library by bio-panning or other techniques, and then scaled-up for expression in a suitable host, characterization and desired applications.

In addition to the above biochemical features, Nbs are highly stable and soluble, resistant to heat, detergents and organic solvents, easily modifiable to introduce additional features or effector functions in the product design, and possess binding ability for less accessible epitopes, *in vivo* tissue penetrance, and low immunogenicity, *etc.* [6-7]. The Nb production technologies are well-perfected, and are highly scalable, and offer low-cost production as additional merit. Moreover, pre-constructed synthetic libraries of huge diversity are excellent starting materials to deliver products specific for the varying epitopes of new infectious agents, within days to weeks, compared to months by the traditional approaches. For these excellent features,Nbs have already been and are increasingly being produced against a vast number of Ags from diverse sources, ranging from those from parasites or their products (viruses, bacteria, fungi, toxins, protozoa, helminthes, *etc.*), host antigens, cancer biomarkers, environment, *etc.* Several Nb-based constructs are undergoing clinical trials in various phases, and promising future bio-therapeutic agents or *in vivo* imaging reagents [6-8].

Applications of Nanobodies

1. Diagnostic Applications

i. *In vitro* Diagnostic Tests

Veterinary diagnostic laboratories with a reliable quality control system for the tests play a central role in optimizing the management and control of infectious diseases, and surveying for epidemiology of these diseases. A plethora of diagnostic tests for infectious animal diseases are currently in use and several are available commercially. Immunodiagnostics tests have employed antisera, later monoclonal and more recently recombinant Ab fragments for detection of antigens or host antibody response. Following categories of 'next generation immunoassays' (NGIAs) are being developed and validated: biosensors for detection of antigens, antibodies or biomarkers, smartphone-based, turbidimetry, paper-based and interferometry-based, multiplex immunoassays based on proteome microarrays, lateral flow biosensors, microbead-based Luminex technology, microfluidics-based 'Lab-on-a-chip', *etc.* [9]. For their excellent features described above, Nbs find place in all of these NGIAs.

A vast number of diagnostically useful Nbs have been produced to target Ags ranging from a number of animal and human viruses (such as rotavirus, H5N1, H7N2, H3N2 influenza viruses, porcine circovirus-2, infectious bursal disease virus, Rift Valley fever virus, HIV-1, foot-and-mouth disease virus, Marburg virus, SARS-CoV, MERS, peste des petit ruminans virus (PPRV), *etc.*), bacteria (such as *Staphylococcus aureus, Escherichia coli, Mycobacterium tuberculosis, Clostridium difficile, Listeria monocytogenes,Brucella melitensis, etc.*), bacterial exotoxins and endotoxins, protozoa (such as *Plasmodium, Trypanosoma spp.*), helminths (such as *Taenia solium, Strongyloidessp.*), insect and snake venoms, cancer biomarkers, enzymes, cytokines, *etc.* [10-24]. These Nbs have been used as reagents in conventional and NGIAs, such as ELISA, Western blotting, lateral flow assays, biosensors, *etc.* For example, camelid Nbs have recently been applied in ELISA methods to detect a wide range of small molecules, including explosive materials (*e.g.*, trinitrotoluene), agents of bioterrorism (botulinum A neurotoxin), toxins (ricin, cholera and staphylococcal enterotoxin B), scorpion toxin, *etc.*

Nb-based constructs include monovalent, bivalent or multivalent formats conjugated to nanoparticles, quantum dots or magnetic particles for use as diagnostic reagents for infectious animal diseases.The biosensors coupled with Nbs (nanoconjugates system) have significantly improved identification of *Staphylococcus aureus* within 10 min. In a recent study, anti-E2 protein of CSFV Nb clones 1 and 2 conjugated with quantum dots (QD)/AF488 were used as two molecular probes for imaging CSFV in cell culture [14]. These Nbs labeled with carboxyl-magnetic beads to construct immunomagnetic nanobeads (IMNBs) were able to capture CSFV virions and recombinant E2 protein, which could be useful for the rapid identification of CSFV during field outbreaks and for research purposes. In another recent study [15], a highly sensitive Nb-based ELISA has been developed for the detection of the Zika virus NSP1 in serum samples. The assay showed sub-nanogram/mL detection limit in the sera and showed excellent reproducibility and accuracy. Similarly, Nbs against H5N1 HPAIV N protein, PPRV Ag, and flaviruses, *viz.*, Rift Valley fever virus, Gnecto and Schmallenberg virus head Ags have been shown to be useful in indirect ELISA [16-17, 22].

Recently, use of the 'sybodies' isolated from synthetic library has been proposed for prophylaxis against COVID-19. Nbs, 8 against S protein and 16 against N protein have been selected from a library in China for rapid Ag detection and treatment of SARS-CoV-2 [20]. In addition, fully human Nbs against five distinct epitopes of SARS-CoV-2 RBD and S1 subunit have also been isolated [21]. Our next-generation sequenced phage library of Indian desert camel Nb clones, however, lacked homology particularly in the CDR3 region to the Nb-72 sequence mentioned above [25-26].

ii. *In vivo* Imaging

The Nb-based formats are more suitable than mAbs for non-invasive imaging techniques. Immuno-PET (Positron Emission Tomography) employing radio-labele dNb can detect specific molecules of interest *in vivo*, such as cancer-specific Ags, and biomarkers associated with various diseases. Nb formats can be efficiently labeled with short-lived γ- or positron-emitting nuclides, rapidly extravasate in tissues to trace their cognate antigens, and the unused products rapidly cleared via the kidneys. Therefore, SPECT (single-photon emission computerized tomography) or PET/CT (positron emission tomography/computerized tomography) imaging can be obtained soon after tracer injection

[7].Cancer biomarkers (such as HER2, VCAM-1) specific Nbs labeled with ^{68}Ga or 99mTc are undergoing clinical trials [8]. Nb could be used as a tool to evaluate novel anti-atherogenic agents. Likewise, various Nbs were developed to monitor disease in mouse models of arthritis. However, their use in veterinary practice is negligible so far and would have a long waiting period before their adoption in veterinary practice, particularly in diagnosis and management of canine malignancies.

2. Passive Protection and Vaccine Quality Control

Immunoprophylaxis is a short-duration once-only preventive approach with the use of antisera for protection against some kinds of toxinoses or other conditions that people in some occupations are more likely exposed. Often the antisera are produced in animal species that differ from the target animal species or humans, their repeated use is constrained due to their inherent immunogenicity for the target host [7]. The same is true for murine mAbs. But, the availability of rDNA-based biotechnological tools permit their modifications to remove easily any kind of constraints, including the immunogenicity (*e.g.*, by humanization), tissue-penetrance and bioavailability (by construction of reduced size Ab fragments, lacking Fc regions), or to introduce any other desired features into the Ag-specific basic Ab format (tailoring and reformatting) [6-7]. Of late, Nbs have been found to be very useful in vaccine quality control studies, particularly for determining vaccine strain purity, antigen mass for the vaccine potency testing, and characterization of the vaccine matching with the field strains.

3. Therapeutics

Serotherapy with antisera, particularly against venoms, bacterial toxins and some viral infections, is an age-old passive protection method in human and veterinary medicine. Monoclonal antibodies have replaced antisera for treatment of some conditions. On similar principles, Nbs promise to replace mAbs with improved outcomes and reduced adverse reactions. Nbs as building blocks have been modified or tailored in order to fulfill the requirements of bioavailability, tissue penetrance, effector functions, *etc.* A broad range of therapeutic applications of Nbs have been proposed by investigators, but only few have reached clinical trials (Table 24.1). One Nb-based anti-vWF drug, designated as Caplacizumab from Sanofi for treating a blood clotting disorder has already been approved by FDA in 2019, and more than 15 are currently undergoing clinical trials and several are in the drug development pipeline [3,7].The Nb clones possessing SARS-CoV-2 neutralizing activity target the trimeric Spike protein and block entry of the virus into host cells have been recently isolated from phage and yeast display libraries, which have broad applications as antivirals and diagnostics in the fight against COVID-19 [27]. Therapeutic applications of the Nb-based products for animal infectious diseases are still a far cry due to several reasons, particularly the costs and technological lag-time constraints. However, with further technological improvements and cost reduction, toxinoses in animals that have been treated traditionally with antisera could become the first possible targets of Nb-based biodrugs in future.

Table 24.1: Therapeutic Applications of Nb-based Biodrugs

Target Disease(s)	*Target Antigens*	*Nb Format*
Acquired Thromotic thrombocytopenic purpura	Von Willebrand factor A	Bivalent, monospecific (Caplacizumab); FDA approved
Cancers: Breast cancer, multiple myeloma, B-cell lymphoma, solid tumors	Tumor-specific Ags, immune checkpoint targets (PD1/PD-L1, CTLA-4), CD19/CD20, CXCR4, *etc.*	Bi- to tetravalent, mono- or bi-specific, CAR-T cells
Viral: respiratory syncytial virus infection, rotavirus diarrhea,	Viral neutralizing epitopes	Bivalent, monospecific
Toxins and venoms: scorpion toxin, botulinum neurotoxin,	Neutralization epitopes	Bivalent, monospecific
Autoimmune diseases: rheumatoid arthritis, systemic lupus erythromatosus, psoriasis	IL6R, IL-17A-IL17F, TNFα, CD28-B7	Bi- or trivalent, mono- or bispecific,
Neurological conditions: Alzheimer	Amyloid-β	Biparatopic
Osteoporosis	RANKL	Trivalent, bispecific

Future Prospects and Conclusions

Nbs have several merits over the conventional Abs for diagnosis and treatment of infectious diseases andtoxinoses. Synthetic Nb libraries are promising tools in the armamentarium to fight against newly emerging pathogens of pandemic potential. Nb-based products can replace mAbs, particularly most-suited for NGIAs, and help strengthen the diagnostic capacity of infectious diseases in low-resource countries. There is an upward trend of the use of Nb-based products in immunodiagnostics and a greater number of animal diseases, new emerging ones in particular, seem to be targeted in future. Veterinary sciences are also expected to draw enormous benefits from these developments in diagnostics, therapeutics and vaccine quality control.

Acknowledgement

I express my gratitude to the organizers of the training at Department of Veterinary Microbiology, CVSc and AH, Jabalpur (MP) for inviting me as expert speaker on this topic.

References

1. OIE. (2020). https: //www.oie.int/en/animal-health-in-the-world/oie-listed-diseases-2020/
2. Bustamante-Córdova L. *et al.* (2018). *Front. Vet. Sci.* **5**: 175.
3. Jovcevska I. and Muyldermans S. (2020). *BioDrugs***34**: 11–26.
4. Chames P and Rothbauer U. (2020). *Antibodies***9**: 6.
5. Arbabi-Ghahroudi M. (2017). *Front. Immunol.***8**: 1589.
6. Muyldermans S. (2020). *The FEBS J.***288**: 2284-2102.
7. Muyldermans S. (2021). *Annu. Rev. Anim. Biosci.* **9**: 401-421.
8. Berland L. *et al.* (2021). *Biomolecules* **11**: 637, doi: 10.3390/biomol11050637.
9. Singh, A. (2017). *Res. and Rev.: J. Immunol. (RRJoI).* **7**(2): 22–43.
10. Yang S. *et al.* (2015). *BMC Biotechnology***15**: 83.
11. Yang S. *et al.* (2015). *J. Virol. Methods***213**: 84–92.
12. Wang D. *et al.* (2015). *BMC Vet. Res.* **11**: 120.
13. Yang S. *et al.* (2018).*Appl. Microbiol. Biotechnol.***102**: 8931–8942.
14. Yang S. *et al.* (2020). *BioMed Res. Internatl.* Article ID 3201630,
15. Delfin-Riela T. *et al.* (2020). *Biomolecules***10**(12): 1652.
16. Cardoso FM. *et al.* (2014). *J. Virol.***88**(15): 8278-8296.
17. Schreur PJW, *et al.* (2020). *eLife***9**: e52716.
18. Weiss RA and Verrips CT. (2019). *Vaccines* **7**(3): 77.
19. Wrapp D. *et al.* (2020). *Cell***181**: 1-12.
20. Jiang S. *et al.* (2020). *Trends Immunol.* **41** (5): 353-357.
21. Custódio TF. *et al.* (2020). *bioRxiv* preprint doi: 10.1101/2020.06.23.165415
22. Kimini E. *et al.* (2021). *Animals***11**: 2206, doi: 10.3390/ani11082206
23. Jangra P and Singh A. (2010). *Asian Pacific J. Trop. Med.* **1**: 1-7.
24. Gupta AK and Singh, A. (2015). *Internatl. J. Infect. Dis.* **45** (Suppl.1): 476-477.
25. Singh A. *et al.* (2020). *Mendeley Data,* V**3**, doi: 10.17632/4rsz3snvk5.3
26. Banerjee S. *et al.* (2020). *Data in Brief***34**: 1-10, doi: 10.1016/j.dib.2020.106663
27. Czajka TF *et al.* (2021). *Trends Microbiol.***29**(3): 195-203.

Chapter 25

Control and Elimination of Rabies

Shrikrishna Isloor

Laboratory Director, KVAFSU-CVA Rabies Diagnostic Laboratory, OiE Reference Laboratory for rabies, Department of Microbiology, Veterinary College, Hebbal, Bengaluru – 560 024, Karanataka
e-mail: kisloor@gmail.com

Rabies is perhaps the most leading viral zoonosis of global importance. The death of the affected host is certain in addition to a major economic burden in developing countries as a result of the high cost of human post exposure prophylaxis, anti rabies immunoglobulins and loss of livestock with special reference to cattle. Most of the countries in Africa and Asia have reported the occurrence of rabies. In India, rabies is endemic and widespread in all parts of the sub-continent except Andaman and Nicobar and Lakshadweep islands. Free ranging dogs play a major role as reservoir and transmit the disease to humans and other domestic animals. Further, most cases are reported in the rural areas where awareness about the disease is less. The situation demands a systematic study on incidence of the disease in animals and creating awareness among the public. The laboratory confirmation of the disease requires collection of brain samples from suspected animals, proper packing, shipping and testing at the laboratory level.

As for the diagnosis of rabies at the laboratory level is concerned, the Direct Fluorescent antibody (DFA) is considered important in view of its sensitivity and specificity. Most of the laboratories across the countries have shifted to DFA from Seller's staining. The KVAFSU-CVA-Crucell Rabies Diagnostic Laboratory was established in 2013 at Veterinary college, Hebbal, Bengaluru with a view to provide a confirmatory diagnosis of rabies in animals and also develop the human resource in India. Later, the laboratory was twinned under World Organization of Animal Health (OIE) Twinning programme in 2016 with the Animal and Plant Health Agency (APHA), Weybridge, UK and Centre for Disease Control and Prevention (CDC), Atlanta, USA as the parent laboratories with the mandate of 'Strengthening diagnosis of rabies in animals in India'. This collaborative effort aimed at (1) Training in all quality assurance systems and health and safety procedures (2) Surveillance in rabies in animals (3) Knowledge transfer and training in antigen detection methods, Virus isolation techniques and Molecular diagnostic methods (4) Proficiency testing and (5) Virus characterization and molecular epidemiology. Various workshops were organized for strengthening the diagnostic capabilities across the country. As a part of this, hands-on training programmes on various aspects such as brain sample collection from Foramen magnum/occipital foramen approach, it's packing, transportation and testing for rabies by Lateral Flow Assay (LFA), Direct Rapid Immunohistochemistry (dRIT), Direct Fluorescent Antibody test (DFA) and Polymerase Chain Reaction (PCR) were organized. In brief, DFA is the most widely used approach which needs fresh specimen: Brain stem and/ or Cerebellum. This test uses rabies virus anti nucleocapsid IgG-FITC conjugate and Evan's blue as the counter stain. The dRIT, a rapid test which is a potential alternative to expensive DFA can be done at bench-top at ambient temperature and is independent of expensive fluorescent microscope, deep freezer and incubator. The dRIT involved incubation of all steps at room temperature itself and requires only a standard light microscope. The test permits rabies diagnosis under the field conditions. The LFA is a rapid immunodiagnostic test useful for ante and Post mortem rabies diagnosis independent of laboratory equipments. As a part of the human resource development in India, Veterinarians from various states were trained on diagnosis of rabies with the motto of capacity building and also laboratory strengthening. Such workshops/training programmes enabled the development of a network of regional rabies diagnostic laboratories across Karnataka, Kerala, Gujarat, Manipur, Andhra Pradesh, Telangana, Maharashtra, Madhyapradesh, Himachal pradesh, Assam, Punjab, Rajasthan, West Bengal, Orissa and Lakshadweep islands. This resulted in the networking of animal rabies diagnostic laboratories in India and increase in the number of samples received at the Rabies Diagnostic Laboratory and also the reporting of the disease. On the other hand, enhancement of immune surveillance through sero-monitoring of rabies vaccinal antibodies in dogs is important to sero monitor anti rabies vaccinal antibodies in dogs and cats for international transportation and evaluating the vaccination programme. The tools used are Rapid Fluorescent Focus Inhibition Test (RFFIT)/Fluorescent Antibody Virus Neutralization Test (FAVN) and Enzyme Linked Immunosorbant Assay (ELISA). Although such sero monitoring has been carried out in some countries, this strategy is not applied in Asian countries. The availability of indigenous, cost effective Recombinant G protein based indirect ELISA in India may further boost the sero monitoring programme in the region.

Mass immunization and seromonitoring have been employed in several countries for control of rabies in dogs. Information, Education and Communication (IEC) based approaches such as workshops, trainings, seminars, conferences creates/increase the awareness about the disease and in turn help in effective control of a disease. Mass awareness programs are also a need in addition to training of veterinarians, strengthening of disease diagnostic laboratories and providing immediate veterinary and medical care in suspected cases. In view of this, the four international organizations namely, the World Health Organization (WHO), The World Organization for Animal Health (OIE), The Food and Agriculture Organization of the United Nations (FAO) and The Global Alliance for Rabies Control (GARC) have united together against rabies with a view to 'Eliminate Dog mediated rabies in humans by 2030'. The three objectives of this 'Rabies Zero by 2030' include 1) Reduce human rabies risk through improved awareness and education, increased access to health care, medicines and vaccines and dog vaccinations 2) Effective policies, governance and guidance through ensuring reliable data to enable effective decision making and 3) Harness multi-stake holder engagement through demonstrating the impact of activities completed under the United against Rabies programme. The success stories of rabies elimination have been documented in Latin America, Bangladesh, Mexico, Phillippines, South Africa, Sri Lanka and united republic of Tanzania. In these countries, two common approaches involving Dog rabies surveillance and Mass dog vaccinations have been successfully carried out. Subsequently, dog rabies surveillance and mass dog anti rabies vaccinations are also planned/initiated by several countries. In view of this, the National Action Plan for Rabies Elimination (NAPRE) has been evolved by the National Centre for Disease Control- NCDC, Directorate General of Health Services, Ministry of Health and Family Welfare, Govt of India, Delhi along with the Department of Animal Husbandry and Dairying (DAHD), Govt. of India and other stake holders. This programme has two components – the Human and Animal Health components. The strategic actions for Animal Health Components aim at estimation of Canine Population, Identification of Rabies Risk Zone, Planning and Implementing Strategic Mass Dog Vaccination Programme, Assessment of Post Vaccination Coverage, Dog Population Management, Promoting Responsible Dog Ownership, Solid Waste Management, Community Involvement and Developing Quarantine Facility. Furthermore, the NCDC is supporting a program for strengthening inter-sectoral co-ordination (ISC) for prevention and control of zoonotic diseases with a mandate of reviewing the current status of rabies control activities in the state and developing a joint state action plan for dog mediated rabies elimination. This program involves various stake holders including State health Departments, representative Medical/ veterinary department from major municipal cooperation, representative disease diagnostic lab., WHO Collaborating Centre on Rabies, National Institute of Mental Health and Neuro Sciences (NIMHANS) Bengaluru, representative State Animal Welfare Board of India (AWBI), Department of Environment,

Forest and Climate Change, Department of urban affairs, Department of waste management/swatch Bharat mission, representative Non Governmental Organization (NGO) working in the field of rabies and representative rabies vaccine manufacturer. Considering these recent developments in India, it is high time that the One Health approach is launched under the NAPRE, laboratories for diagnosis of animal rabies in India are strengthened, Special surveys for dog population enumeration are carried out, mass dog vaccination campaigns are initiated throughout the country including both urban and rural areas, Government organizations, NGOs, INGOS are involved in mass dog vaccination and 'stake holders' in the government are sensitized about rabies control in animals with a goal to achieve " Dog mediated rabies control by 2030".

The recent designation of KVAFSU-CVA Rabies Diagnostic Laboratory, an ISO 17025/2017 accredited Laboratory at the Department of Microbiology, Veterinary college, Hebbal, Bengaluru as the OIE Reference Laboratory for rabies is a big strength to India in further harmonizing the animal rabies surveillance with special reference to dogs and implementing Mass dog vaccinations in the region. In this nationally important task, it is necessary that all the stake holders including the community are well coordinated.

Index

A

Adjuvants 7
African swine fever 99
Aluminium hydroxide gel vaccine 33
Alum precipitated vaccine 32
American Veterinary Medical Association 115
Animal disease diagnosis 73
Animals 165
Antibiotic resistance patterns 175, 178
Antibody virus neutralization Test 206
Antigenic diversity 94
Antigen-lateral flow devices 79
Antigen retrieval 50
Antimicrobial profiling 162
Antimicrobial resistance 83
Antimicrobials 84
Aquatic animal diseases 39
Arsenic 166
Attenuted vaccines 62

B

Bacterial identification 162
Benign footrot 141
Big cats 145
Biofilm forming 83
Biofilm producing bacteria 87
Biological inactivation 59
Biological sampling 147
Biosafety laboratory design 67
Bovine herpesvirus-1 infections 153
Bovines 1
Bovine viral diarrhea 14
Brucella 1
Brucellosis 2

C

Cadmium 170
Canine distemper virus 121
Challenges 9
Chemical inactivation 59
Clinical signs 101, 123

Clinical techniques 41

Coccidial vaccine 94

Components 3

Control program 189

Culture techniques 108

Cytokine 94

D

Department of Animal Husbandry and Dairying 207

Diagnostic applications 199

Diagnostic methods 125

Diagnostic microbiology 157

Dipstick test 188

Direct rapid immunohistochemistry 206

Disease diagnosis 47

Disease investigation 147

DNA vaccines 5, 35

E

Efficacy test 25

Eimeria oocyst 93

Emerging diseases 1

Endogenous peroxidase activity 51

Enterobacteriaceae family 83

Enzyme-linked immunosorbent assay (ELISA) 17, 126, 206

Extended-spectrum-Beta-lactamase 85

F

Final product testing 23

Flow cytometry 18

Food and Agriculture Organization of the United Nations (FAO) 115, 207

Foot and Mouth Disease (FMD) 1, 14

Footrot 139

Formaldehyde estimation 27

G

Genital form 155

Ghost vaccine 35

Global alliance for rabies control 207

Gram-negative bacteria 85

H

Haemagglutination 135

Haemorrhagic septicaemia vaccines 31

Heavy metal toxicity 165

Hemorrhagic septicemia 2

Histopathological investigation 147

Holistic approach 111

I

Immunocompromised hosts 2

Immunofluorescent antibody test 18

Immunohistochemical staining 51

Immunohistochemistry 17, 47

Immunohistochemistry protocol 52

Immunological diagnostic methods 161

Immunological investigations 148

Inactivated antigen testing 23

Inactivated vaccine 25, 63

Indirect fluorescent antibody test 126

Infectious diseases 48, 197

Intermediate footrot 142

In vitro diagnostic tests 199

In vivo imaging 200

K

Killed/inactivated vaccines 4

Killed vaccines 58

L

Laboratory biosafety 67

Laboratory practices 70

Lateral flow assay 18, 206
Lateral flow chromatographic immunoassay 161
Lead 168
Liposome 59
Live antigen testing 23
Live attenuated vaccines 4, 59
Live rDNA products 28
Livestock 83
Live vaccine 25, 34
Loop mediated isothermal amplification (LAMP) assay 188

M

Madlin Darby bovine kidney 154
Malignant catarrhal fever 14
Manufacturing process 28
Mass vaccination 62
Mercury 171
Methicillin-resistant *Staphylococcus aureus* 175
Microbial diagnostic and identification kits 159
Microbiological and epidemiological aspect 176
Moisture content 28
Molecular diagnosis 149
Molecular epidemiology 185
Molecular techniques 109
Monoclonal antibody 13, 16
Morphological characterization 158
mRNA vaccines 6
Mycoplasma infections 107

N

Nanobodies 197
National action plan for rabies elimination 207
National animal disease control programme 2
New generation vaccines 31
Next-generation sequencing 149
Nobel approaches 83
Nucleic acid vaccines 60

O

OIE 115
Oil adjuvant vaccine 33
One health 111

P

Parvo virus infection 133
Passive protection 201
Pathogenesis 123, 167
Pathological investigations 148
Peritoneal macrophages 14
Personal protective equipment 71
Peste-des-Petits-ruminants 183
Phacochoerus aethiopicus 100
Postmortem 101
Potency testing 26
Poultry 57
Poultry coccidiosis 91
PPRV nucleic acid 188
Prophylaxis 121, 127

Q

Quality control 21, 22

R

Rabies 205
Rapid and sensitive technologies 162
Rapid fluorescent focus inhibition test 206
Raw materials 23
Real-time RT-PCR 188
Recombinant vaccine 59, 94

Record keeping 28
Respiratory form 155
Rinderpest 1, 14, 21
Risk factors 177
Routine laboratory procedures 41

S

Safety equipment 70
Safety test 24
Sandwich ELISA 188
Serological techniques 108
Spleen cells 15
Stability tests 27
Sterility test 24
Subunit vaccines 5, 35

T

Technical considerations 49
Testing 21
Therapeutics 201
Transmission 85, 122

V

Vaccine assessment 26
Vaccine delivery methods 8
Vaccine quality control 201
Vaccine technologies 1, 4
Veterinary vaccines 21
Viral zoonosis 205
Virulent footrot 141
Virus cultivation 154
Virus structure 153

W

Western immunoblotting 17
Wildlife 118
World Animal Health Organization 197
World Health Organization (WHO) 115, 207
World Organization for Animal Health 207

Z

Zoonotic diseases 119